Lecture Notes in Electrical Engineering

Volume 118

T0142123

For further volumes:
http://www.springer.com/series/7818

Ahmed Abdelgawad · Magdy Bayoumi

Resource-Aware Data Fusion Algorithms for Wireless Sensor Networks

 Springer

Ahmed Abdelgawad
54 Lavoie Drive
  Essex Junction
VT 05452, USA
ama1916@cacs.louisiana.edu

Magdy Bayoumi
University of Louisiana
  at Lafayette
Lafayette, Louisiana, USA
mab@cacs.louisiana.edu

ISSN 1876-1100   e-ISSN 1876-1119
ISBN 978-1-4899-8706-8   ISBN 978-1-4614-1350-9 (eBook)
DOI 10.1007/978-1-4614-1350-9
Springer New York Dordrecht Heidelberg London

Printed on acid-free paper

Springer is part of Springer Science+Business Media (www.springer.com)

Preface

WSN (Wireless Sensors Networks) is intended to be deployed in environments where sensors can be exposed to circumstances that might interfere with measurements provided. Such circumstances include strong variations of pressure, temperature, radiation, and electromagnetic noise. Thus, measurements may be imprecise in such scenarios. Data fusion is used to overcome sensor failures, technological limitations, and spatial and temporal coverage problems.

Not many books addressed the real life problem in WSN applications. In this book, we are proposing real implementation of data fusion algorithms; taking into consideration the resource constrains of WSN. In addition, we are introducing some real applications, as case study, in the industry.

The data fusion can be implemented in both centralized and distributed systems. In the centralized fusion case, we propose four algorithms to be implemented in WSN. As a case study, we propose a remote monitoring framework for sand production in pipelines. Our goal is to introduce a reliable and accurate sand monitoring system. The framework combines two modules: a Wireless Sensor Data Acquisition (WSDA) module and a Central Data Fusion (CDF) module. The CDF module is implemented using four different proposed fusion methods; Fuzzy Art (FA), Maximum Likelihood Estimator (MLE), Moving Average Filter (MAF), and Kalman Filter (KF). All the fusion methods are evaluated throughout simulation and experimental results. The results show that FA, MLE and MAF methods are very optimistic, to be implemented in WSN, but Kalman filter algorithm does not lend itself for easy implementation; this is because it involves many matrix multiplications, divisions, and inversions. The computational complexity of the centralized KF is not scalable in terms of the network size. Thus, we propose to implement the Kalman filter in a distributed fashion. The proposed DKF is based on a fast polynomial filter to accelerate distributed average consensus. The idea is to apply a polynomial filter on the network matrix that will shape its spectrum in order to increase the convergence rate by minimizing its second largest eigenvalue. Fast convergence can contribute to significant energy savings. In order to implement the DKF in WSN, more power saving is needed. Since multiplication is the atomic operation of Kalman filter, saving power at the multiplication level can significantly impact the energy consumption of the DKF. This work also proposes a novel light-weight and low-power multiplication algorithm. Experimental results show that the TelosB mote can run DKF with up to seven neighbors.

This book is based on Abdelgawad PHD dissertation. The work presented was carried out through a large scale research project titled UCoMS (Ubiquitous Computing and Monitoring System) supported by DoE and State of Louisiana.

We appreciate the support, the project team, and the working environment of UCoMS. The VLSI group infrastructure, stimulating and challenging environment, and the weakly presentation and discussion have been an asset to the presented work.

Abdelgawad offers all praise to the almighty God, Allah, the Most Gracious, and the Most Merciful for his blessings bestowed upon him and for giving him the strength to achieve what he has accomplished in his life. Abdelgawad dedicates this book to his family which has played an important role in his life and study. Their support and encouragement has made this book a reality. He would like to thank his mother for her prayers, love, and faith in him. Ahmed's deepest appreciation goes to his lovely wife, Dalia Aboelfadl, his precious daughter, Salma, his handsome son, Mohamed, and his little son, Ali for their unlimited encouragement, sacrifices, and for being by his side.

Bayoumi would like to dedicate this book to his smart, energetic, and dedicated students.

Lafayette, Louisiana   Ahmed Abdelgawad
                                                                        Magdy Bayoumi

Contents

List of Figures

List of Tables

List of Abbreviations

ADC	Analog-to-digital converter
ASIC	Application specific integrated circuits
CCA	Clear channel assessment
CDF	Central data fusion
CKF	Central Kalman filters
CMOS	Complementary metal–oxide–semiconductor
CoD	Conditioning and digitizing
CPU	Central processing unit
CSMA	Carrier sense multiple access
CTS	Clear to send
DC	Direct current
DEM	Decentralized expectation maximization
DKF	Distributer Kalman filter
DoD	Department of Defense
DP	Differential pressure
DPM	Dynamic power management
DSP	Digital signal processors
DVS	Dynamic voltage scaling
EEPROM	Electrically erasable programmable read-only memory
FA	Fuzzy art
FIR	Finite impulse response
FPGA	Field programmable gate array
GUI	Graphical user interface
HCI	Human computer interaction
I/O	Input/output
IIR	Impulse response filter

IP	Internet protocol
ISM	Industrial scientific and medical
JDL	Joint directors of laboratories
KCF	Kalman-consensus filters
KF	Kalman filter
LAN	Local area network
LCD	Liquid crystal display
LMI	Linear matrix inequality
LPL	Low power listening
MaC	Management and control
MAC	Medium access control
MAC	Multiply accumulate unit
MAF	Moving average filter
Max	Maximum
MIMO	Multiple-input and multiple-output
ML	Maximum likelihood
MLE	Maximum likelihood estimator
NiCd	Nickel-cadmium
Nimh	Nickel metal hydride
NiZn	Nickel-zinc
NP	Nondeterministic polynomial
P2P	Peer-to-peer
QoS	Quality of service
RAM	Random-access memory
ReT	Receiving and transmission
RF	Radio frequency
RISC	Reduced instruction set computing
RTS	Request to send
SPI	Serial peripheral interface
TCP	Transmission control protocol
TDMA	Time division multiple access
UAV	Unmanned aerial vehicle
WSDA	Wireless sensor data acquisition
WSN	Wireless sensor network

Chapter 1
Introduction

Abstract A Wireless Sensor Network (WSN) is a network comprised of numerous small autonomous sensor nodes called motes. It combines a broad range of networking, hardware, software, and programming methodologies. Each node is a computer with attached sensors that can process and exchange sensed data, as well as communicates wirelessly among them to complete various tasks. Sensors attached to this node allow them to sense various phenomena within the surroundings. WSN has received momentous attention in recent years because of its titanic potential in applications. In this chapter, we introduced many applications of WSN, explained the sensor node evaluation metrics, brought in the sensor network architecture, and finally we discussed the WSN's challenges and constraints.

A Wireless Sensor Network (WSN) is a network comprised of numerous small autonomous sensor nodes called motes. It combines a broad range of networking, hardware, software, and programming methodologies. Each node is a computer with attached sensors that can process and exchange sensed data, as well as communicates wirelessly among them to complete various tasks. Sensors attached to this node allow them to sense various phenomena within the surroundings. Characteristics of a wireless sensor network include the capability to make autonomous actions based on surrounding observations. Motes need to be self-organizing, self-regulated, self-repairing, and programmable. The mote technology is rather constrained in order to provide a low-cost, reusable deployment into varying environments. Although each node is able to deal with a variety of jobs, it has many limitations as well. Memory capacity of a node is limited. Furthermore, most of the nodes currently available in the market are battery-operated; hence they have a limited life-time. These limitations are a major factor and must be addressed when designing and implementing a WSN. As an example, a routing algorithm for WSN must be memory and energy efficient. Since radio transmissions use a significant amount of energy, researchers seek ways to reduce radio communication as much as possible. However, when more information is stored and more computation is done to reduce the communication costs, energy consumption of the processor and

A. Abdelgawad and M. Bayoumi, *Resource-Aware Data Fusion Algorithms*
for Wireless Sensor Networks, Lecture Notes in Electrical Engineering 118,
DOI 10.1007/978-1-4614-1350-9_1, © Springer Science+Business Media, LLC 2012

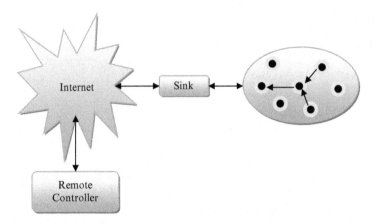

Fig. 1.1   Sensor network architecture

memory components are becoming an important issue. Design choices have to be made, and these also depend on the intended application. Figure 1.1 shows the sensor network architecture.

The power of wireless sensor networks lies in the capability to install large numbers of tiny nodes that configure themselves. Usage scenarios for these devices range from real-time tracking, to ubiquitous computing environments, to monitoring of environmental conditions. While often referred to as wireless sensor networks, they can also control actuators that extend control from cyberspace into the world. The simplest application of wireless sensor network technology is to monitor remote environments for low frequency data trends. For example, a chemical plant could be simply monitored for leaks by hundreds of sensors that automatically form a wireless interconnection network and instantly report the detection of any chemical leaks. Unlike traditional wired systems, installation costs would be minimal. Instead of having to install thousands of feet of wire routed through protective conduit, installers simply have to place quarter-sized devices [1].

In addition to radically reducing the installation costs, the wireless sensor network has the ability to dynamically adjust with the changing of the environments. Adjustment mechanisms can respond to changes in network topologies or can cause the network to shift between radically different modes of operation. For example, the same embedded network performing leak monitoring in a chemical factory might be reconfigured into a network designed to localize the source of a leak and track the flow of poisonous gases. The network could then direct workers to the safest route for emergency evacuation.

Unlike traditional wireless devices, wireless sensor nodes do not need to communicate directly with the nearest high-power control base station, but only with their local peers. Instead of relying on a pre-deployed communications, each individual sensor becomes part of the overall communications. Peer-to-peer networking protocols provide a mesh-like interconnect to transfer data between the thousands of tiny embedded devices in a multi-hop fashion. The flexible mesh architectures envisioned animatedly adapt to support introduction of new nodes or expand to cover a larger geographic area. As well, the system can automatically adjust to compensate for node failures. The vision of mesh networking is based on strength in numbers. Unlike cell phone schemes that deny service when too many phones are active in a small area, the interconnection of a wireless sensor network only grows faster as nodes are added. As long as there is enough density, a single network of nodes can grow to cover limitless area. With each node having a communication range of 150 ft and costing less that \$1 a sensor network that surrounded the equator of the earth will cost less than \$1 M.

The wireless sensor network architecture includes both a hardware platform and an operating system designed specifically to meet with the needs of wireless sensor networks. TinyOS is a component-based operating system designed to run in resource constrained wireless devices. It provides an extremely efficient communication primitives and fine-grained concurrency mechanisms to application and protocol developers. A key concept in TinyOS is the use of event-based programming in conjunction with a highly professional component model. TinyOS enables system-wide optimization by providing a tense coupling between hardware and software, as well as flexible mechanisms for building application-specific modules. TinyOS has been designed to run on a generalized architecture, where a single Central Processing Unit (CPU) is shared between application and protocol processing.

1.1   Wireless Sensor Network Applications

The applications for WSNs are varied, typically involving some kind of monitoring, tracking, or controlling. Specific applications include habitat monitoring, object tracking, nuclear reactor control, fire detection, and traffic monitoring. In a typical application, a WSN is scattered in a region where it is meant to collect data through its sensor nodes [2].

1. Area monitoring: Area monitoring is a regular application of WSNs. In area monitoring, the WSN is deployed over an area where some phenomenon is to be monitored. For example, a large quantity of sensor nodes could be deployed over a battlefield to sense enemy intrusion instead of using landmines. When the sensors sense the event being monitored, i.e., pressure, light, electro-magnetic field, sound, vibration, heat, etc., the event needs to be reported to one of the base stations, which can do appropriate action, e.g., send a message on the internet or to a satellite. Depending on the application, different objective functions will require different data-propagation strategies, depending on things such as need for real-time response, redundancy of the data, which can be done via data aggregation and information fusion techniques, need for security, etc. [3].

2. Environmental data collection: An environmental data collection application is one where a research scientist wants to collect several sensor readings from a set of points in an environment over a period of time in order to detect styles and interdependencies. This scientist would want to collect data from hundreds of points extending throughout the area and then analyze the data later offline. The scientist would be interested in collecting data over several weeks, months or years in order to look for long-term and seasonal trends. For the data to be significant it would have to be collected at regular intervals and the nodes would remain at known locations. At the network level, the environmental data collection application is differentiated by having a large number of nodes continually sensing and transmitting data back to a set of base stations that store the data using traditional methods. These networks generally need very low data rates and extremely long lifetimes. In typical usage scenario, the nodes will be evenly distributed over an outdoor environment. This distance between nearby nodes will be minimal yet the distance across the entire network will be significant. After deployment, the nodes must discover the topology of the network first and then estimate optimal routing strategies. The routing strategy can then be used to direct data to a central collection points. In environmental monitoring applications, it is not necessary that the nodes develop the optimal routing strategies on their own. Instead, it may be possible to determine the optimal routing topology outside of the network and then communicate the necessary information to the nodes as required. This is possible because the physical topology of the network is relatively stable. While the time variant nature of Radio Frequency (RF) communication may cause connectivity between two nodes to be alternating, the overall topology of the network will be relatively stable. Environmental data collection applications typically use tree-based routing topologies where each routing tree is rooted at high-capability nodes that sink data. Data is periodically transmitted from child node to parent node up the tree-structure until it reaches the sink. With tree-based data collection each node is in charge of forwarding the data of all its descendants. Nodes with a large number of descendants transmit significantly more data than leaf nodes. These nodes can quickly become energy bottlenecks. The most essential characteristics of the environmental monitoring requirements are long lifetime, precise synchronization, low data rates and relatively static topologies. Additionally it is not important that the data be transmitted in real-time back to the central collection point. The data transmissions can be delayed inside the network as necessary in order to improve network efficiency [4].

3. Landfill ground well level monitoring and pump counter: Wireless sensor networks can be used to evaluate and monitor the water levels within all ground wells in the landfill site and monitor leachate accumulation and removal. A wireless device and submersible pressure transmitter monitors the leachate level. The sensor information is wirelessly transmitted to a central data logging system to store the level data, make calculations, or inform personnel when a service vehicle is needed at a specific well. It is typical for leachate removal pumps to be installed with a totalizing counter mounted at the top of the well to

monitor the pump cycles and to determine the total volume of leachate removed from the well. For most current installations, this counter is read physically. Instead of physically collecting the pump count data, wireless devices can send data from the pumps back to a central control location to save time and get rid of errors. The control system uses this count information to determine when the pump is in process, to determine leachate extraction volume, and to schedule maintenance on the pump [5].

4. Security monitoring: Security monitoring networks are composed of nodes that are placed at fixed locations all over an environment that continually monitor one or more sensors to detect an irregularity. A difference between security monitoring and environmental monitoring is that security networks are not actually collecting any data. This has a significant impact on the optimal network architecture. Each node has to regularly check the status of its sensors but it only has to transmit a data report when there is a security violation. The immediate and reliable communication of alarm messages is the main system requirement. These are reported by exception networks. Additionally, it is essential that it is confirmed that each node is still there and functioning. If a node were to be disabled or fail, it would represent a security violation that should be reported. For security monitoring applications, the network must be configured so that nodes are in charge of confirming the status of each other.

5. Vehicle detection: If traffic controls are implemented on the whole traffic network, transportation capability could be maximized. The controls are greatly dependent on the data from traffic surveillance systems, which have a high installation and repairs costs. In view of this, researchers offer a very attractive, low-cost solution, which applies wireless sensor networks for traffic surveillance. A traffic surveillance system requires four components: a sensor to catch the signals made by vehicles, a processor to process the sensed data, a communication unit to transfer the processed data to the base station, and an energy source. Thanks to sensor technology, all of these components could now be integrated into a single tiny device.

6. Agriculture: Using wireless sensor networks within the agricultural industry are more and more common. Gravity-fed water systems can be monitored using pressure transmitters to monitor water tank levels, pumps can be controlled using wireless I/O devices, and water use can be measured and wirelessly transmitted back to a central control center for billing. Irrigation automation enables more professional water use and reduces waste.

7. Windrow composting: Composting is the aerobic decomposition of biodegradable organic matter to produce compost, a nutrient-rich mulch of organic soil formed using food, wood, manure, and/or other organic material. One of the key methods of composting involves using windrows. To ensure efficient and useful composting, the temperatures of the windrows must be measured and logged frequently. With accurate temperature measurements, facility managers can determine the best time to turn the windrows for quicker compost production. Manually collecting data is time wasting, cannot be done frequently, and may expose the person collecting the data to harmful pathogens. Automatically collecting the data

and wirelessly transmitting the data back to a centralized location allows composting temperatures to be continually recorded and logged, reducing the time needed to complete a composting cycle, improving efficiency, and minimizing human exposure and potential risk. An industrial wireless I/O device mounted on a stake with two thermocouples, each at different depths, can automatically monitor the temperature at two depths within a compost windrow or stack. Temperature sensor readings are wirelessly transmitted back to the host system for data collection, analysis, and logging. Because the temperatures are measured and recorded continuously, the composting rows can be turned as soon as the temperature reaches the best point. Continuously monitoring the temperature may also provide an early warning to possible fire hazards by notifying personnel when temperatures exceed recommended ranges.

8. Node tracking: There are many situations where one would like to track the location of an important asset or personnel. Current inventory control systems attempt to track objects by recording the last checkpoint that an object passed through. However, with these systems it is not possible to determine the current location of an object. For example, UPS tracks every shipment by scanning it with a barcode whenever it passes through a routing center. The system breaks down when objects do not flow from checkpoint to checkpoint. In typical work environments it is not practical to expect objects to be continually passed through checkpoints. With wireless sensor networks, object can be tracked by simply tagging it with a small sensor node. The sensor node will be tracked as it moves through a field of sensor nodes that are deployed in the environment at known locations. Instead of sensing environmental data, these nodes will be deployed to sense the RF messages of the nodes attached to various objects. The nodes can be used as active tags that announce the existence of a device. A database can be used to record the location of tracked objects relative to the set of nodes at known locations.

9. Greenhouse monitoring: Wireless sensor networks are also used to control the temperature and humidity levels inside greenhouses. When the temperature and humidity goes down below specific levels, the greenhouse manager must be notified via e-mail or cell phone text message, or host systems can trigger misting systems, turn on fans, open vents, or control a wide variety of system responses. Because some wireless sensor networks are easy to install, they are also easy to move as the needs of the application change [6].

1.2 Sensor Node Evaluation Metrics

The key evaluation metrics for wireless sensor nodes are power, flexibility, robustness, security, communication, computation, Time Synchronization, size and cost. Their importance is discussed below [7].

1. Power: To meet the multi-year application requirements, individual sensor nodes must be incredibly low-power. This ultra-low-power operation can only be

completed by combining both low-power hardware components and low duty-cycle operation procedures. During active operation, radio communication will constitute a major fraction of the node's total energy budget. Algorithms and protocols must be developed to reduce radio activity whenever possible. This can be achieved by using localized computation to reduce the streams of data being generated by sensors and through application-specific protocols. For example, events from multiple sensor nodes can be combined together by a local group of nodes before transmitting a single result across the sensor network.

2. Flexibility: The broad range of usage scenarios being considered means that the node architecture must be flexible and adaptive. Each application scenario will require a slightly different mix of lifetime, sample rate, response time, and in-network processing. Wireless sensor network architecture must be flexible enough to accommodate a wide range of applications. Additionally, for cost reasons each device will have only the hardware and software it really needs for a given application. The architecture must make it easy to assemble just the right set of software and hardware components. Thus, these devices require an abnormal degree of hardware and software modularity while simultaneously maintaining efficiency.

3. Robustness: In order to support the lifetime requirements demanded, each node must be created to be as robust as possible. In a typical deployment, hundreds of nodes will have to work in harmony for years. To accomplish this, the system must be constructed so that it can tolerate and adjust to individual node failure. Additionally, each node must be designed to be as robust as possible. System modularity is a powerful tool that can be used to develop a robust system. By dividing system functionality into isolated sub-pieces, each function can be fully tested in isolation earlier to combining them into a complete application. To facilitate this, system components should be as independent as possible and have interfaces that are narrow, in order to prevent unexpected interactions. In addition to increasing the system's robustness to node failure, a wireless sensor network must also be robust to external interference. As these networks will often coexist with other wireless systems, they need the talent to adjust their behavior consequently. The robustness of wireless links to external interference can be really increased through the use of multi-channel and spread spectrum radios. It is common for facilities to have existing wireless devices that work on one or more frequencies. The talent to avoid congested frequencies is essential in order to guarantee a successful deployment.

4. Security: In order to meet the application level security requirements, the individual nodes must be able to perform complex encrypting and validation algorithms. Wireless data communication is easily susceptible to interception. The only method to maintain data carried by these networks confidential and authentic is to encrypt all data transmissions. The CPU must be able to perform the required cryptographic operations itself or with the help of included cryptographic accelerators. In addition to securing all data transmission, the nodes themselves must secure the data that they have. While they will not have large amounts of application data stored internally, they will have to store secret

encryption keys used in the network. If these keys are exposed, the security of the network could collapse. To provide true security, it must be difficult to extract the encryption keys from any node.

5. Communication: A key evaluation metric for any wireless sensor network is its communication rate, power consumption, and range. While we have made the argument that the coverage of the network is not limited by the transmission range of the individual nodes, the transmission range does have a significant impact on the minimal acceptable node density. If nodes are placed too far apart it may not be possible to create an interconnected network or one with enough redundancy to maintain a high level of reliability. Most applications have natural node densities that correspond to the granularity of sensing that is desired. If the radio communications range demands a higher node density, extra nodes must be added to the system in to increase node density to a tolerable level. The communication rate also has a major impact on node performance. Higher communication rates turn into the ability to achieve higher effective sampling rates and lower network power consumption. As bit rates increase, transmissions take less time and therefore potentially require less energy. However, an increase in radio bit rate is often accompanied by an increase in radio power consumption. All things being equal, a higher transmission bit rate will result in higher system performance.

6. Computation: The two most computationally intensive operations for a wireless sensor node are the in-network data processing and the management of the low-level wireless communication protocols. There are strict real-time requirements associated with both communication and sensing. As data is arriving over the network, the CPU must concurrently control the radio and record/decode the incoming data. Higher communication rates required faster computation. The same is true for processing being performed on sensor data. Analog sensors can produce thousands of samples per second. Common sensor processing operations include digital filtering, threshold detection, averaging, correlation and spectral analysis. It may even be necessary to perform a real-time FFT on incoming data in order to detect a high-level event. In addition to being able to process, refine and discard sensor readings, it can be beneficial to combine data with neighboring sensors before transmission across a network. Just as complex sensor waveforms can be reduced to key events, the results from multiple nodes can be synthesized together. This in-network processing requires additional computational resources. Beyond that, the application data processing can consume an arbitrary amount of computation depending on the calculations being performed.

7. Time synchronization: In order to support time correlated sensor readings and low-duty cycle operation of data collection application, nodes must be able to keep precise time synchronization with other members of the network. Nodes need to sleep and awake together so that they can once in a while communicate. Errors in the timing mechanism will create inefficiencies that result in increased duty cycles. In distributed systems, clocks drift apart over time due to inaccuracies in timekeeping mechanisms. Depending on temperature, voltage,

and humidity, time keeping oscillators operate at slightly different frequencies. High-precision synchronization mechanisms must be provided to continually compensate for these inaccuracies.

8. Size and cost: The physical size and cost of each individual sensor node has a considerable and direct impact on the ease and cost of deployment. Total cost of ownership and initial deployment cost are two key factors that will drive the implementation of wireless sensor network technologies. In data collection networks, researchers will often be operating off of a fixed budget. Their primary goal will be to collect data from as many locations as possible without exceeding their fixed budget. A reduction in per-node cost will result in the ability to purchase more nodes, deploy a collection network with higher density, and collect more data. Physical size also impacts the ease of network deployment. Smaller nodes can be placed in more locations and used in more scenarios. In the node tracking scenario, smaller, lower cost nodes will result in the ability to track more objects.

1.3 Sensor Network Architecture

The main components of a sensor node are microcontroller, transceiver, external memory, power source and one or more sensors as shown in Fig. 1.2.

1. Microcontroller: Microcontroller processes data and controls the functionality of other components in the sensor node. Other alternatives that can be used as a controller are: General purpose desktop microprocessor, Digital Signal Processors (DSP), Field Programmable Gate Array (FPGA) and Application-Specific Integrated Circuit (ASIC). Microcontrollers are the most suitable choice for a sensor node. Each of the four choices has its own advantages and disadvantages. Microcontrollers are the best choices for embedded systems. Because of their flexibility to connect to other devices, programmable, power consumption is less, as these devices can go into a sleep mode and part of the controller can be active. In a general purpose microprocessor the power consumption is more than the microcontroller, therefore it is not a suitable choice for sensor node. Digital Signal Processors are suitable for broadband wireless

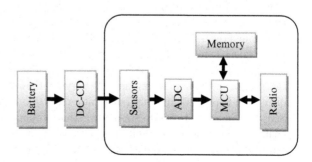

Fig. 1.2 Sensor node architecture

communication. But in WSNs, the wireless communication should be reserved i.e., simpler, easier to process modulation and signal processing tasks of actual sensing of data is less complicated. Therefore the advantages of DSP are not of that much importance to wireless sensor node. FPGA can be reprogrammed and reconfigured according to requirements, but it takes time and energy. Therefore, FPGA is not advisable. ASIC is specialized processor designed for a given application. ASIC provides the functionality in the form of hardware, but microcontrollers provide it through software [8].

2. Transceiver: Transceiver makes use of Industrial, Scientific and Medical (ISM) band which gives free radio, massive spectrum allocation and universal availability. The various choices of wireless transmission media are RF, optical communication and infrared. Optical communication requires less energy, but needs line-of-sight for communication and is also sensitive to atmospheric conditions. Infrared like optical communication, needs no antenna but is limited in its broadcasting capacity. RF based communication is the most relevant that fits to most of the WSN applications. WSN uses the communication frequencies between about 433 MHz and 2.4 GHz. The functionality of both transmitter and receiver, combined into a single device know as transceivers, are used in sensor nodes. Transceivers lack a unique identifier. The operational states are transmit, receive, idle and sleep. Current generation radios have a built-in state machine that performs this operation automatically.

3. External memory: From an energy point of view, the most relevant kinds of memory are on-chip memory of a microcontroller and flash memory. Flash memories are used due to their cost and storage capacity. Memory requirements are very much application-dependent.

4. Power source: Power consumption in the sensor node is for the communication, data processing and sensing. More energy is required for data communication in the sensor node. Energy overhead is less for data processing and sensing. The energy cost of transmitting 1 Kb a distance of 300 ft is approximately the same as that for the executing 3 million instructions by 100 million instructions per second/W processor. Power is stored either in batteries or capacitors. Batteries are the main source of power supply for sensor nodes. Namely, the two types of batteries used are chargeable and non-rechargeable. They are also classified according to electrochemical material used for electrode such as NiCd (nickel–cadmium), NiZn (nickel–zinc), Nimh (nickel metal hydride), and Lithium-Ion. Current sensors are developed which are able to renew their energy from solar, vibration, or temperature. Two major power saving policies used are Dynamic Power Management (DPM) and Dynamic Voltage Scaling (DVS). DPM takes care of shutting down parts of the sensor node which are not currently used or active. DVS scheme varies the power levels depending on the non-deterministic workload. By varying the voltage along with the frequency, it is possible to obtain quadratic reduction in power consumption [9].

5. Sensors: Sensors are hardware devices that produce measurable response to a change in a physical condition like temperature, humidity and pressure. Sensors sense or measure physical data of the area to be monitored. The continual analog

signal sensed by the sensors is digitized by an Analog-to-Digital Converter (ADC) and sent to controllers for further processing. Characteristics and requirements of the sensor node should be small size, consume extremely low energy, operate in high volumetric densities, be autonomous and operate unattended, and be adaptive to the environment. As wireless sensor nodes are microelectronic sensor device, can only be equipped with a limited power source of less than 0.5 Ah and 1.2 V. Sensors are classified into three categories:

(a) Passive sensors: They sense the data without actually manipulating the environment by active probing. They are self powered, i.e., energy is needed only to amplify their analog signal. There is no notion of direction involved in these measurements. A typical example is the camera.
(b) Active sensors: These groups of sensors actively probe the environment; for example, a sonar or radar sensor or some type of seismic sensor, which generate shock waves by small explosions.
(c) Omni-directional sensors: Each sensor node has a certain area of coverage for which it can reliably and accurately report the particular quantity that it is observing.

6. MAC: A Medium Access Control (MAC) protocol coordinates actions over a shared communication channel. The most commonly used solutions are contention-based. One general contention-based strategy is for a node which has a message to transmit to test the channel to see if it is busy, if not busy then it transmits; otherwise it waits and tries again later. After colliding, nodes wait a random amount of time trying to avoid re-colliding. If two or more nodes transmit at the same time there is a collision and all the nodes colliding try to transmit again later. Many wireless MAC protocols also have a dozen modes where nodes not involved with sending or receiving a packet in a given timeframe go into sleep mode to save energy. An effective MAC protocol for wireless sensor networks must avoid collisions, consume little power, be implemented with a small code size and memory requirements, be efficient for a single application, and be tolerant to changing radio frequency and networking conditions. One example of a good MAC protocol for wireless sensor networks is B-MAC [10]. B-MAC is highly configurable and can be implemented with a small code and memory size. It has an interface that allows choosing various functionality and only that functionality as needed by a particular application. B-MAC consists of four main parts: Clear Channel Assessment (CCA), packet backoff, link layer acts, and low power listening. For CCA, B-MAC uses a weighted moving average of samples when the channel is idle in order to assess the background noise and better be able to detect valid packets and collisions. The packet backoff time is configurable and is chosen from a linear range as opposed to an exponential backoff scheme typically used in other distributed systems. This reduces delay and works because of the typical communication patterns found in a wireless sensor network. B-MAC also supports a packet by packet link layer acknowledgement. In this way only important packets need pay the extra cost. A low power listening scheme is employed where a node cycles

between awake and sleep cycles. While awake it listens for a long enough preamble to assess if it needs to stay awake or can return to sleep mode. This scheme saves significant amounts of energy. Many MAC protocols use a Request To Send (RTS) and Clear To Send (CTS) style of interaction. This works well for ad hoc mesh networks where packet sizes are large (thousands of bytes). However, the overhead of RTS-CTS packets to set up a packet transmission is not acceptable in wireless sensor networks where packet sizes are on the order of 50 bytes. B-MAC, therefore, does not use a RTS-CTS scheme. Another example of a good MAC protocol for wireless sensor networks is Z-MAC [11]. Z-MAC is a hybrid MAC protocol for wireless sensor networks. It combines the strengths of Time Division Multiple Access (TDMA) and Carrier Sense Multiple Access (CSMA) while offsetting their weaknesses. Unlike TDMA, where a node is allowed to transmit only during its own assigned slots, a node can transmit in both its own time slots and slots assigned to other nodes. Owners of the current time slot always have priority in accessing the channel over non-owners. Therefore, under low contention where not all owners have data to send, non-owners can steal time slots from owners. This has the effect of switching between CSMA and TDMA depending on contention. Z-MAC is robust to topology changes and clock synchronization errors; in the worst case its performance falls back to that of CSMA. Synchronized protocols, such as S-MAC [12] and T-MAC [13], negotiate a schedule that specifies when nodes are awake and asleep within a frame. Specifying the time when nodes must be awake in order to communicate reduces the time and energy wasted in idle listening. Asynchronous protocols such as WiseMAC [14], rely on Low Power Listening (LPL), also called preamble sampling. Standard MAC protocols developed for duty-cycled WSNs employ an extended preamble and preamble sampling. While this "low power listening" approach is simple, asynchronous, and energy-efficient, the long preamble introduces excess latency at each hop, is suboptimal in terms of energy consumption, and suffers from excess energy consumption at receivers. X-MAC [15] proposes solutions to each of these problems by employing a shortened preamble approach that retains the advantages of low power listening, namely low power communication, simplicity and a decoupling of transmitter and receiver sleep schedules.

1.4   Wireless Sensor Network Challenges

In this section we present some of the major WSNs' challenges. Challenges for WSNs may be categorized as follows: resource constraints, platform heterogeneity, dynamic network topology and mixed traffic.

1. Resource constraints: As in WSNs, sensor nodes are usually low-cost, low-power, small devices that are equipped with only limited data processing capability, transmission rate, battery energy, and memory. For example, the MICAz mote

from Crossbow is based on the Atmel ATmega128L 8-bit microcontroller that provides only up to 8 MHz clock frequency, 128-KB flash program memory and 4-KB Electrically Erasable Programmable Read-Only Memory (EEPROM); the transmit data rate is limited to 250 kbps. Due to the limitation on transmission power, the available bandwidth and the radio range of the wireless channel are often limited. In particular, energy conservation is critically important for extending the lifetime of the network, because it is often unfeasible or undesirable to recharge or replace the batteries attached to sensor nodes once they are deployed. In the presence of resource constraints, the network Quality of Service (QoS) may suffer from the unavailability of computing and/or communication resources. For instance, a number of nodes that want to transmit messages over the same WSN have to compete for the limited bandwidth that the network is able to provide. As a consequence, some data transmissions will possibly experience large delays, resulting in low level of QoS. Due to the limited memory size, data packets may be dropped before the nodes successfully send them to the destination. Therefore, it is of critical importance to use the available resources in WSNs in a very efficient way.

2. Platform heterogeneity: WSNs are designed using different technologies and with different goals; they are different from each other in many aspects such as computing/communication capabilities, functionality, and number. In a large-scale system of systems, the hardware and networking technologies used in the WSNs may differ from one subsystem to another. This is true because of the lack of relevant standards dedicated to WSNs and hence commercially available products often have disparate features. This platform heterogeneity makes it very difficult to make full use of the resources available in the integrated system. Consequently, resource efficiency cannot be maximized in many situations. In addition, the platform heterogeneity also makes it challenging to achieve real-time and reliable communication between different nodes.

3. Dynamic network topology: Unlike LANs, where nodes are typically stationary, the WSNs may be mobile. In fact, node mobility is an intrinsic nature of many applications such as, among others, intelligent transportation, assisted living, urban warfare, planetary exploration, and animal control. During runtime, new sensor nodes may be added; the state of a node is possibly changed to or from sleeping mode by the employed power management mechanism; some nodes may even die due to exhausted battery energy. All of these factors may potentially cause the network topologies of WSNs to change dynamically. Dealing with the inherent dynamics of WSNs requires QoS mechanisms to work in dynamic and even unpredictable environments. In this context, QoS adaptation becomes necessary; that is, WSNs must be adaptive and flexible at runtime with respect to changes in available resources. For example, when an intermediate node dies, the network should still be able to guarantee real-time and reliable communication by exploiting appropriate protocols and algorithms.

4. Mixed traffic: Diverse applications may need to share the same WSN, inducing both periodic and aperiodic data. This feature will become increasingly evident as the scale of WSNs grows. Some sensors may be used to create the

measurements of certain physical variables in a periodic manner for the purpose of monitoring and/or control. Meanwhile, some others may be deployed to detect critical events. For instance, in a smart home, some sensors are used to sense the temperature and lighting, while some others are responsible for reporting events like the entering or leaving of a person. Furthermore, disparate sensors for different kinds of physical variables, e.g., temperature, humidity, location, and speed, generate traffic flows with different characteristics (e.g. message size and sampling rate).

Bibliography

1. S. Petersen and S. Carlsen, "Wireless Sensor Networks: Introduction to Installation and Integration on an Offshore Oil & Gas Platform," in *Proceeding of the 19th Australian Conference on Software Engineering,* Washington DC, USA, March 2008, pp. 53–53.
2. M. Galetzka, J. Haufe, M. Lindig, U. Eichler, and P. Schneider, "Challenges of simulating robust wireless sensor network applications in building automation environments," in *Proceeding of the IEEE Conference on Emerging Technologies and Factory Automation,* Bilbao, Spain, September 2010, pp. 1–8.
3. W. You-Chiun, H. Yao-Yu, and T. Yu-Chee, "Multiresolution Spatial and Temporal Coding in a Wireless Sensor Network for Long-Term Monitoring Applications," *IEEE Transactions on Computers,* vol. 58, pp. 827–838, April 2009.
4. P. M. Glatz, L. B. Hormann, C. Steger, and R. Weiss, "Implementing autonomous network coding for wireless sensor network applications," in *Proceeding of the 18th International Conference on Telecommunications,* Graz, Austria, June 2011, pp. 9–14.
5. S. A. Butt, P. Sayyah, and L. Lavagno, "Model-based hardware/software synthesis for wireless sensor network applications," in *Proceeding of the Saudi International Electronics, Communications and Photonics Conference,* Riyadh, Saudi Arabia, April 2011, pp. 1–6.
6. P. A. Morreale, "Wireless Sensor Network Applications in Urban Telehealth," in *21st International Conference on Advanced Information Networking and Applications Workshops,* Niagara Falls, Ontario, Canada, May 2007, pp. 810–814.
7. L. Barolli, T. Yang, G. Mino, A. Durresi, F. Xhafa, and M. Takizawa, "Performance Evaluation of Wireless Sensor Networks for Mobile Sensor Nodes Considering Goodput and Depletion Metrics," in *Proceeding of the 9th IEEE International Symposium on Parallel and Distributed Processing with Applications,* Dresden, Germany, August, 2011, pp. 63–68.
8. W. Fenhua, L. Fang, W. Zhiliang, and G. Jingjing, "Wireless sensor network architecture design and implementation," in *Proceeding of the 3rd IEEE International Conference on Broadband Network and Multimedia Technology,* Beijing, China, October 2010, pp. 1068–1073.
9. D. Benhaddou, M. Balakrishnan, and X. Yuan, "Remote Healthcare Monitoring System Architecture using Sensor Networks," in *IEEE Region 5 Conference,* Fayetteville, Arkansas, USA, April 2008, pp. 1–6.
10. J. Polastre, J. Hill, and D. Culler, "Versatile low power media access for wireless sensor networks," in *Proceeding of the 2nd ACM International Conference on Embedded Networked Sensor Systems,* Baltimore, MD, USA, November 2004, pp. 95–107.
11. I. Rhee, A. Warrier, M. Aia, and J. Min, "ZMAC: a Hybrid MAC for Wireless Sensor Networks," in *Proceeding of the SenSys,* San Diego, California, USA, November 2005, pp. 56–61.

12. Y. Wei, J. Heidemann, and D. Estrin, "Medium access control with coordinated adaptive sleeping for wireless sensor networks," *IEEE/ACM Transactions on Networking,* vol. 12, pp. 493–506, June 2004.
13. T. V. Dam and K. Langendoen, "An adaptive energy-efficient mac protocol for wireless sensor networks," in *Proceeding of the 1st ACM Conf. on Embedded Networked Sensor Systems,* Los Angeles, California, USA, November 2003, pp. 171–180.
14. A. El-Hoiydi and J. D. Decotignie, "WiseMAC: an ultra low power MAC protocol for the downlink of infrastructure wireless sensor networks," in *Proceeding of the 9th International Symposium on Computers and Communications,* Alexandria, Egypt, June 2004, pp. 244–251.
15. E. A. M. Buettner, G. Yee, and R. Han, "X-mac: A short preamble mac protocol for duty-cycled wireless sensor networks," in *Proceeding of the 4th ACM Conference on Embedded Sensor Systems,* New York, NY, USA, April 2006, pp. 307–320.

Chapter 2
Data Fusion in WSN

Abstract WSN is intended to be deployed in environments where sensors can be exposed to circumstances that might interfere with measurements provided. Such circumstances include strong variations of pressure, temperature, radiation, and electromagnetic noise. Thus, measurements may be imprecise in such scenarios. Data fusion is used to overcome sensor failures, technological limitations, and spatial and temporal coverage problems. Data fusion is generally defined as the use of techniques that combine data from multiple sources and gather this information in order to achieve inferences, which will be more efficient and potentially more accurate than if they were achieved by means of a single source. The term efficient, in this case, can mean more reliable delivery of accurate information, more complete, and more dependable. The data fusion can be implemented in both centralized and distributed systems. In a centralized system, all raw sensor data would be sent to one node, and the data fusion would all occur at the same location. In a distributed system, the different fusion modules would be implemented on distributed components. Data fusion occurs at each node using its own data and data from the neighbors. This chapter briefly discusses the data fusion and a comprehensive survey of the existing data fusion techniques, methods and algorithms.

2.1 Introduction

A Wireless Sensor Network (WSN) may be designed with different objectives. It may be designed to gather and process data from the environment in order to have a better understanding of the behavior of the monitored area. It may also be designed to watch an environment for the occurrence of a set of possible events, thus the proper action may be taken whenever needed. A fundamental issue in WSN is the way to process the collected data. In this situation, data fusion arises as a discipline that is concerned with how data collected by sensors can be processed to increase the significance of such a mass of data [1]. Thus, data fusion can be defined as the combination of multiple sources to obtain improved data i.e., cheaper, greater

A. Abdelgawad and M. Bayoumi, *Resource-Aware Data Fusion Algorithms*
for Wireless Sensor Networks, Lecture Notes in Electrical Engineering 118,
DOI 10.1007/978-1-4614-1350-9_2, © Springer Science+Business Media, LLC 2012

quality, or greater relevance. Data fusion is commonly used in detection and classification tasks in different application domains, such as military applications and robotics [2]. Within the WSN domain, simple aggregation techniques i.e., maximum, minimum, and average have been used to reduce the overall data traffic to save energy [3, 4]. Additionally, data fusion techniques have been applied to WSNs to improve location estimates of sensor nodes, detect routing failures, and collect link statistics for routing protocols [5].

WSN is intended to be deployed in environments where sensors can be exposed to circumstances that might interfere with measurements provided. Such circumstances include strong variations of pressure and temperature, radiation and electromagnetic noise. Thus, measurements may be imprecise in such scenarios. Even when environmental conditions are ideal, sensors may not give perfect measurements. Basically, a sensor is a measurement device, and vagueness is usually associated with its observation. Such imprecision represents the imperfections of the technology and methods used to measure a physical incident. Failures are not an exception in WSN. For example, consider a WSN that monitors a jungle to detect an event, such as fire or the presence of an animal. Sensor nodes can be destroyed by fire, animals, or even human beings; they might present manufacturing problems; and they might stop working due to a lack of energy. Each node that becomes inoperable might compromise the overall perception and/ or the communication capability of the network. Here, perception ability is equivalent to the exposure concept. Both spatial and temporal coverage also pose limitations to WSN. The sensing capability of a node is restricted to a limited area. For example, a thermometer in a room reports the temperature near the device but it might not represent fairly the overall temperature inside the room. Spatial coverage in WSN has been explored in different scenarios, such as node scheduling, target tracking, and sensor placement. Temporal coverage can be understood as the ability to fulfill the network purpose during its lifetime. For example, in a WSN for event detection, temporal coverage aims at assuring that no relevant event will be missed because there was no sensor perceiving the region at the specific time the event occurred. Thus, temporal coverage depends on the sensor's sampling rate, node's duty cycle, and communication delays. To overcome sensor failures, technological limitations, and spatial and temporal coverage problems, three properties must be ensured:

1. Cooperation.
2. Redundancy
3. Complementarily

Usually, the area of interest can only be completely covered by the use of several sensor nodes, each cooperating with a partial view of the scene; and data fusion can be used to create the complete view from the pieces provided by each node. Redundancy makes the WSN less vulnerable to failure of a single node, and overlapping measurements can be fused to obtain more precise data. Complementarily can be achieved by using sensors that observe different properties of the environment; data fusion can be used to combine complementary data so the resultant data allows inferences that might be not possible to be obtained from the individual measurements, e.g., angle and distance of an imminent threat can be fused to obtain its position. Due to redundancy and cooperation properties, WSN is often composed of a large number of sensor nodes posing a new scalability challenge caused by possible collisions and transmissions of redundant data. Regarding the energy restrictions, communication should be reduced to increase the lifetime of the sensor nodes. Hence, data fusion is also important to reduce the overall communication load in the network by avoiding the transmission of redundant messages. In addition, any task in the network that handles signals or needs to make inferences can potentially use data fusion. Data fusion should be considered a critical step in designing a wireless sensor network. The reason is that data fusion can be used to extend the network lifetime and is commonly used to fulfill the application objectives, such as event detection, target tracking, and decision making. Hence, careless data fusion may result in waste of resources and misleading assessments. Therefore, we must be aware of possible limitations of data fusion to avoid blundering situations. Because of the resource rationalization needs of WSN, data processing is commonly implemented as in-network algorithms. Hence, data fusion should be performed in a distributed fashion to extend the network lifetime. Even so, we must be aware of the limitations of distributed implementations of data fusion. Thus, regarding the communication load, a centralized fusion system may outperform a distributed one. The reason is that centralized fusion has a global knowledge in the sense that all measured data is available, whereas distributed fusion is incremental and localized since it fuses measurements provided by a set of neighbor nodes and the result might be further fused by intermediate nodes until a sink node is reached. Such a drawback of decentralized fusion might often be present in WSN wherein, due to resource limitations, distributed and localized algorithms are preferable to centralized ones.

Data fusion has established itself as an independent research area over the last decades, but a general formal theoretical framework to describe data fusion systems is still missing. One reason for this is the huge number of disparate research areas that utilize and illustrate some form of data fusion in their context of theory. For example, the concept of data or feature fusion, which forms together with classifier and decision fusion the three main divisions of fusion levels, initially occurred in multi-sensor processing. By now several other research fields found its application useful. Besides the more classical data fusion approaches in statistics, control, robotics, computer vision, geosciences and remote sensing, artificial intelligence, and digital image/signal processing, the data retrieval community discovered some years ago its power in combining multiple data sources.

2.2 Information Fusion, Sensor Fusion, and Data Fusion

Several different terms have been used to illustrate the aspects regarding the fusion subject, e.g. information fusion, sensor fusion, and data fusion. The expressions related to systems, applications, methods, architectures, and theories about the fusion of data from multiple sources are not unified yet. Different terms have been adopted, usually associated with particular aspects that characterize the fusion i.e., sensor fusion is commonly used to specify that sensors provide the data being fused. Despite the theoretical issues about the difference between information and data, the terms information fusion and data fusion are usually accepted as overall terms. Many definitions of data fusion have been provided along the years, most of them were found in military and remote sensing fields. The data fusion work group of the Joint Directors of Laboratories (JDL) organized an effort to define a dictionary with some terms of reference for data fusion [6]. They define data fusion as a multilevel process dealing with the automatic detection, estimation, association, correlation, and combination of data and data from several sources. The JDL data fusion model deals with quality improvement. Hall defines data fusion as a combination of data from multiple sensors to accomplish improved accuracy and more specific inferences that could be achieved by the use of a single sensor alone [7]. All the previous definitions are focused on means, methods and sensors. Wald in [8] changes the attention of fuse data to the used framework. He defines data fusion as a formal framework in which is expressed means and tools for the alliance of data originating from different sources. He considers data taken from the same source at different instants as separate sources. For WSN, data can be fused with at least two objectives: accuracy improvement and energy saving.

Multisensor integration is another expression used in computer vision and industrial automation. Luo [9] defines multisensor integration as a synergistic use of data provided by multiple sensory devices to help in the accomplishment of a task by a system. However, multisensor fusion deals with the combination of different sources of sensory data into one representational format during any stage in the integration process. Multisensor integration is a broader term than multisensor fusion. It makes clear how the fused data is used by the whole system to interact with the environment. However, it might suggest that only sensory data is used in the fusion and integration processes.

The term data aggregation term has become popular in the wireless sensor network community as a synonym for information fusion [10]. Data aggregation comprises the collection of raw data from pervasive data sources, the flexible, programmable composition of the raw data into less voluminous refined data, and the timely delivery of the refined data to data consumers. Aggregation is the ability to summarize data i.e., the amount of data is reduced. However, for applications that require original and accurate measurements, such summarization may represent an accuracy loss [11]. Although many applications might be interested only in summarized data, we cannot always state whether or not the summarized data is more precise than the original data set. Because of that, the use of data aggregation as a general term should be avoided because it also refers to one example of data fusion, which is summarization. Figure 2.1 shows the relationship among the concepts of multisensor/sensor fusion, multisensor integration, data aggregation, information fusion, and data fusion. Here, we understand that both terms, information fusion and data fusion, can be used with the

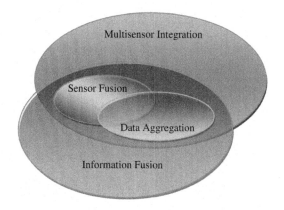

Fig. 2.1 The relationship among the fusion terms: multisensor/sensor fusion, multisensor integration, data aggregation, information fusion and data fusion

same meaning. Multisensor/sensor fusion is the subset that operates with sensory sources. Data aggregation defines another subset of information fusion that means to reduce the data volume, which can manipulate any type of information/data, including sensory data. Thus, multisensor integration is a slightly different term in the sense that it applies information fusion to make inferences using sensory devices and associated information to interact with the environment. Thus, multisensor/sensor fusion is fully contained in the intersection of multisensor integration and information/data fusion.

2.3 Data Fusion Classification

Data fusion can be classified based on several features. Relationships among the input data can be used to divide data fusion into:

1. Cooperative data
2. Redundant data
3. Complementary data.

The abstraction level of the manipulated data during the fusion process can be used to distinguish among fusion processes as:

1. Measurement
2. Signal
3. Feature
4. Decision

Another general classification considers the abstraction level, and it makes explicit the abstraction level of the input and output of a fusion process.

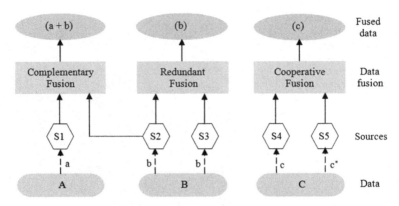

Fig. 2.2 Types of data fusion based on the relationship among the sources

2.3.1   Classification Based on Relationship Among the Sources

Data fusion can be classified, according to the relationship among the sources [9]. Thus, data fusion can be:

1. Complementary: Data provided by the sources represents different portions of a broader scene; data fusion can be applied to obtain a piece of data that is more complete. In Fig. 2.2, sources S1 and S2 provide different pieces of data (a and b) that can be fused to achieve a complete data (a + b) composed of non-redundant pieces a and b that refer to different parts of the environment. In general, complementary fusion searches for completeness by compounding new data from different pieces. Hoover [12] applies complementary fusion by using several cameras to observe different parts of the environment; then the video streams are fused into an occupancy map that is used to guide a mobile robot. An example of complementary fusion consists in fusing data from sensor nodes, e.g., a sample from the sensor field, into a feature map that describes the whole sensor field [13].
2. Redundant: If two or more independent sources provide the same piece of data, these pieces can be fused to increase the associated confidence. Sources S2 and S3 in Fig. 2.2 provide the same data (b). S2 and S3 are fused to obtain more accurate data (b). Redundant fusion might be used to increase the reliability, accuracy, and confidence of the data. In WSN, redundant fusion can provide high quality data and prevent sensor nodes from transmitting redundant data.
3. Cooperative: Independent sources are cooperative when the data provided by them is fused into new data that represents the reality. Sources S4 and S5 in Fig. 2.2, provide different data, c and c*, that are fused into (c), which better describes the scene compared to c and c* individually. A traditional example of

cooperative fusion is the computation of a target location based on angle and distance data. Cooperative fusion should be carefully applied since the resultant data is subject to the inaccuracies and imperfections of all participating sources.

2.3.2 Classification Based on Levels of Abstraction

Luo in [14] applied four levels of abstraction to classify data fusion:

1. Signal level fusion: It deals with single sensors and can be used in real-time applications or as an intermediate step for further fusions.
2. Pixel level fusion: It operates on images and can be used to improve image-processing tasks.
3. Feature level fusion: Deals with features or attributes extracted from signals or images, such as speed and shape.
4. Symbol level fusion: Data is a symbol that represents a decision, and it is also referred to a decision level.

In general, the feature and symbol fusions are used in object recognition applications. This classification presents some disadvantages and is not suitable for all data fusion applications. First, both images and signals are considered raw data and are usually provided by sensors, so they should be included in the same class. Second, raw data may not be only from sensors, because data fusion systems might also fuse data provided by databases or human interaction. Third, it proposes that a fusion process cannot deal with all levels at the same time.

According to the level of abstraction of the manipulated data, data fusion can be classified into four categories:

1. Low-level fusion: Raw data are provided as inputs and combined into new data that are more accurate than the individual inputs. Polastre in [15] gave an example of low-level fusion by applying a moving average filter to estimate ambient noise and determine whether or not the communication channel is clear.
2. Medium-level fusion: Features and attributes of an entity are fused to obtain a feature map that may be used for other tasks. It is also known as feature/attribute level fusion.
3. High-level fusion: It is known as symbol or decision level fusion. It takes decisions or symbolic representations as input and combines them to obtain a more confident and/or a global decision. An example of high-level fusion is the Bayesian approach for binary event detection proposed by Krishnamachari in [16] that detects and corrects measurement faults.
4. Multilevel fusion: Fusion process encompasses data of different abstraction levels and both input and output of fusion can be of any level. For example, a measurement is fused with a feature to provide a decision.

2.3.3   Classification Based on Input and Output

Dasarathy introduced another classification that considers the abstraction level. Data fusion processes are categorized based on the level of abstraction of the input and output data [17]. He identifies five categories:

1. Data in – data out (DAI-DAO): In this class, data fusion deals with raw data and the result is also raw data, possibly more accurate or reliable.
2. Data in – feature out (DAI-FEO): Data fusion uses raw data from sources to extract features or attributes that describe an entity. Entity here means any object, situation, or world abstraction.
3. Feature in – feature out (FEI-FEO): It works on a set of features to improve/ refine a feature, or extract new ones.
4. Feature in – decision out (FEI-DEO): Data fusion takes a set of features of an entity generating a symbolic representation or a decision.
5. Decision in – decision out (DEI-DEO): Decisions can be fused in order to obtain new decisions or give emphasis on previous ones.

In comparison to the classification presented before, this classification can be seen as an extension of the earlier one with a finer granularity where DAI-DAO corresponds to Low Level Fusion, FEI-FEO to Medium Level Fusion, DEI-DEO to High Level Fusion, DAI-FEO and FEI-DEO are included in Multi-level Fusion.

2.4   Data Fusion: Techniques, Methods, and Algorithms

Techniques, methods, and algorithms used to fuse data can be classified based on several criteria, such as the data abstraction level, parameters, mathematical foundation, purpose, and type of data. Data fusion can be performed with different objectives such as inference, estimation, feature maps, aggregation, abstract sensors, classification, and compression.

2.4.1   Inference

Inference method is applied in decision fusion. The decision is taken based on the knowledge of the perceived situation. At this point, inference refers to the transition from one likely true proposition to another, which its truth is believed to result from the previous one. Classical inference methods are based on the Bayesian inference and the Dempster-Shafer belief about accumulation theory.

1. Bayesian inference: Data fusion based on Bayesian Inference provides a formalism to merge evidence according to rules of probability theory. The uncertainty is represented in terms of conditional probabilities describing the belief, and it can assume values in the [0, 1] interval, where 0 is the absolute disbelief and 1 is the absolute belief. Within the WSN domain, Bayesian inference has been used to solve the localization problem. Sichitiu in [18] uses the Bayesian inference to process data from a mobile beacon and determine the most likely geographical location of each node, as an alternative of finding a unique point for each node location.

2. Dempster-Shafer inference: The Dempster-Shafer inference is based on the Dempster-Shafer belief accumulation, which is a mathematical theory introduced by Dempster [19] and Shafer [20] that generalizes the Bayesian theory. It deals with beliefs or mass functions just as Bayes' rule does with probabilities. The Dempster-Shafer theory introduced a formalism that can be used for incomplete knowledge representation and evidence combination. Pinto discussed in-network implementations of the Dempster-Shafer and the Bayesian inference in such a way that event detection and data routing are combined into a single algorithm [21]. By using a WSN composed of Unmanned Aerial Vehicle (UAV) as sensor nodes, Yu uses the Dempster-Shafer inference to build dynamic operational pictures of battlefields for situation evaluation. However, the particular challenges of in-network fusion in such a mobile network are not evaluated [22].

3. Fuzzy logic: Fuzzy logic generalizes probability and, therefore, is able to deal with approximate reasoning to draw conclusions from imprecise premises. Each quantitative input is fuzzyfied by a membership function. The fuzzy rules of an inference system generate fuzzy outputs which, in turn, are defuzzyfied by a set of output rules. This structure has been successfully used in real world situations that defy exact modeling, from rice cookers to complex control systems. Gupta uses fuzzy reasoning for deciding the best cluster-heads in a WSN [23].

4. Neural networks: Neural Networks represent an alternative to Bayesian and Dempster-Shafer theories, being used by classification and recognition tasks in the data fusion domain. A key feature of neural networks is the capability of learning from examples of input/output pairs in a supervised fashion. For that reason, neural networks can be used in learning systems while fuzzy logic is used to control its learning rate. Neural networks have been applied to data fusion mainly for automatic target recognition using multiple complementary sensors.

5. Semantic data fusion: In semantic data fusion, raw sensor data is processed so that nodes exchange only the resulting semantic interpretations. The semantic abstraction allows a WSN to optimize its resource utilization when storing, collecting, and processing data. Semantic data fusion usually comprises two phases: pattern matching and knowledge-base construction. Friedlander [24] introduced the concept of semantic data fusion, which was applied for target classification.

2.4.2 Estimation

Estimation method was inherited from control theory and used the laws of probability to compute a process state vector from a measurement vector or a sequence of measurement vectors. We present, in this section, the estimation methods known as: Least Squares, Maximum Likelihood (ML), Moving Average filter, Kalman filter, and Particle filter.

1. Least squares: Least Squares method is a mathematical optimization technique that searches for a function that best fits a set of input measurements. This is accomplished by minimizing the sum of the square error between points generated by the function and the input measurements. The Least Squares method is suitable when the parameter to be estimated is considered fixed. Least Square method does not assume any prior probability.
2. Maximum likelihood: Estimation methods based on Likelihood are suitable when the state being estimated is not the outcome of a random variable. Xiao proposes a distributed and localized Maximum Likelihood that is robust to the unreliable communication links of WSN. In this method, every node computes a local unbiased estimate that converges towards the global Maximum Likelihood solution [25]. Xiao further extended this method to support asynchronous and timely delivered measurements, i.e., measurements taken at different time steps that happen asynchronously in the network. Other distributed implementations of ML for WSN include the Decentralized Expectation Maximization (DEM) algorithm and the local Maximum Likelihood estimator that relax the requirement of sharing all the data [26].
3. Moving average filter: Moving average filter is broadly adopted in digital signal processing (DSP) solutions as it is simple to understand and use. Moreover, this filter is optimal for reducing random white noise while retaining a sharp step response. This is the reason that makes the moving average the major filter for processing encoded signals in the time domain. As the name suggests, this filter computes the arithmetic mean of a number of input measurements to produce each point of the output signal. Yang uses the Moving Average filter on target locations to reduce errors of tracking applications in WSNs [27].
4. Kalman filter: Kalman filter is a very popular fusion method. It was originally proposed in 1960 by Kalman [28] and it has been extensively studied since then. Kalman filter is used to fuse low-level redundant data. If a linear model can describe the system and the error can be modeled as Gaussian noise, the Kalman filter recursively retrieves statistically optimal estimates. On the other hand, to deal with non-linear dynamics and non-linear measurement models other methods should be adopted. In WSN, we can find schemes to approximate distributed Kalman filter, in which the solution is computed based on reaching an average consensus among sensor nodes [29].
5. Particle filter: The Particle filter is a recursive implementation of a statistical signal processing known as sequential Monte Carlo methods. Although

Kalman filter is a classical approach for state estimation, particle filters represent an alternative for applications with non-Gaussian noise, especially when computational power is rather cheap and sampling rate is slow. The particles are propagated over time, sequentially combining, sampling, and resampling steps. At each time step, the resampling is used to discard some particles, increasing the relevance of regions with high posterior probability. Target tracking is currently the principal research problem wherein particle filters have been used.

2.5   Data Fusion: Architectures and Models

Many architectures and models have been introduced to serve as guidelines to design data fusion systems. These models evolved from data-based models to role-based models. These models are useful for guiding the specification, proposal, and usage of data fusion in WSN. Some of these models, such as the JDL and Frankel-Bedworth, provide a systemic view of data fusion, whereas others, such as the Intelligent Cycle and the Boyd Control Loop, provide a task view of data fusion.

2.5.1   Data-Based Models

Models proposed to design data fusion systems can be centered on the abstraction of the data generated during fusion. This section introduces the models that specify their stages based on the abstraction levels of data manipulated by the fusion system [1].

1. JDL model: JDL is a well-known model in the fusion research area. It was originally proposed by the U.S. Joint Directors of Laboratories (JDL) and the U.S. Department of Defense (DoD). The model consists of five processing levels, an associated database, and a data bus connecting all components as shown in Fig. 2.3. Its components are described as follows:

   - Sources: It is responsible for providing the input data and can be sensors, a prior knowledge, databases, or human input.
   - Database management system: It supports the maintenance of the data used and provided by the data fusion system. This is a critical function, as it supposedly handles a large and varied amount of data. In WSNs, this function might be simplified to fit the sensors' restrictions of resources.
   - Human computer interaction (HCI): It is a mechanism that allows human input, such as commands and queries, and the notification of fusion results through alarms, displays, graphics, and sounds. Commonly, human interaction with WSNs occurs through the query-based interfaces.

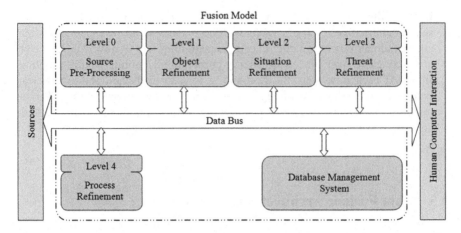

Fig. 2.3 The JDL model

- Level 0 (source preprocessing): It is also referred to as process alignment, this
  level aims to reduce the processing load by allocating data to appropriate
  processes and selecting appropriate sources.
- Level 1 (object refinement): It converts the data into a consistent structure.
  Source localization, and therefore, all tracking algorithms are in Level 1,
  since they transform different types of data, such as images, angles, and
  acoustic data, into a target location.
- Level 2 (situation refinement): It attempts to provide a contextual description
  of the relationship between objects and observed events. It uses a prior
  knowledge and environmental data to identify a situation.
- Level 3 (threat refinement): It estimates the current situation, projecting it in
  the future to identify possible threats, vulnerabilities, and opportunities for
  operations. This is a difficult task because it deals with computation
  complexities and enemies intent assessment.
- Level 4 (process refinement): It is responsible for monitoring the system
  performance and allocating the sources according to the specified goals.
  This function may be outside the domain of specific data fusion functions.

2. Dasarathy model: The Dasarathy model [17] is a fine-grained data-centered
   model in which the elements of data fusion are specified based on their inputs
   and outputs. It is known also as Data-Feature-Decision (DFD) [17]. Figure 2.4
   depicts the DFD model.

The primary input is raw data and the main output is a decision. The components
responsible for the several fusion stages are the elements DAI-DAO, DAI-FEO,
FEIFEO, FEI-DEO and DEI-DEO, described before. The DFD model is successful
in specifying the main types of fusion regarding their input and output data. For this
reason it is also used to classify data fusion. In contrast to the JDL model, the DFD
model does not provide a systemic view; instead it provides a fine-grained way to

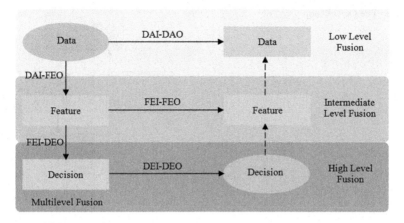

Fig. 2.4 The DFD model

specify fusion tasks by means of the expected input and output data. Therefore, the DFD model is useful for specifying and designing fusion algorithms in WSNs with different purposes such as ambient noise estimation.

2.5.2   Activity-Based Models

Some models are specified based on the activities that must be performed by the data fusion system. The activities and their correct sequence of execution, in such models, are explicitly specified.

1. Boyd control loop: The Boyd Control Loop is a cyclic model composed of four stages. It is known also as the Observe, Orient, Decide, Act (OODA) Loop. This model is a representation of the classic decision-support mechanism of military data systems, and because such systems are strongly coupled with fusion systems, the OODA loop has been used to design data fusion systems. The stages of the OODA loop define the major activities related to the fusion process as shown in Fig. 2.5, which are:

   - Observe: Data gathering from the available sources. It corresponds to level 0 of the JDL model.
   - Orient: Gathered data is fused to obtain an interpretation of the current situation. It encompasses levels 1, 2, and 3 of the JDL model.
   - Decide: Specify an action plan in response to the understanding of the situation. It matches level 4 of JDL model.
   - Act: The plan is executed. It is not dealt by the JDL model.

   The OODA loop is a wide model that allows the specification and visualization of the system tasks in an ample way. It allows the modeling of the main tasks of a system. The OODA fails to provide a proper representation of specific tasks of a data fusion system.

Fig. 2.5 The OODA loop

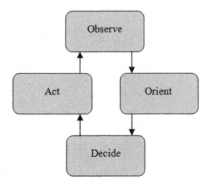

Fig. 2.6 The intelligence
cycle

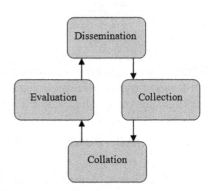

2. Intelligence cycle: The intelligence process is a four-stage cycle, which is called
   Intelligence Cycle. Figure 2.6, shows the process of developing raw data into
   finished intelligence used in decision-making and action. The activities of the
   Intelligence Cycle are:

   - Collection: Raw data is collected from the environment. It matches level 0 of
     the JDL model.
   - Collation: Collected data is compared, analyzed, and correlated. Irrelevant
     and unreliable data is discarded. Includes level 1 of the JDL model.
   - Evaluation: Collated data is fused and analyzed. It comprises levels 2 and 3 of
     the JDL model.
   - Dissemination: Fusion results are delivered to users who utilize the fused data
     to produce decisions and actions in response to the detected situation.
     It corresponds to level 4 of the JDL model.

   The Intelligence Cycle does not make explicit the planning (Decide) and the
   execution (Act) phases, which are most likely included in the Evaluation and
   Dissemination phases. The OODA and Intelligence Cycle are general and can be
   employed in any application domain. They do not fulfill the specific aspects of
   the fusion domain demanding, thus, experience and expertise to model fine-
   grained fusion tasks.

Fig. 2.7 Omnibus model

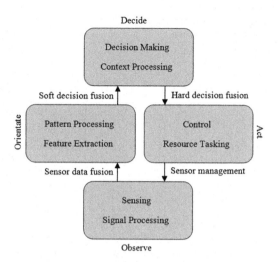

3. Omnibus model: The Omnibus model organizes the stages of a data fusion system in a cyclic sequence, just as the Intelligence Cycle and the OODA loop do [30]. The Omnibus model should be applied during the design phase of a data fusion system. Initially, it should be used to model the framework providing a general perception of the system. Then, the model can be used to design the subtasks, providing a fine-grained understanding of the system. Figure 2.7 shows the Omnibus model. The Omnibus model was originally proposed to deal with data collected by sensor devices. Some modifications can be suggested to make it more broad and suitable for other data fusion systems such as:

- Sensing and signal processing can be replaced by data gathering and data preprocessing, respectively.
- Sensor data fusion should be stated as raw data fusion.
- Instead of Sensor management we should adopt source management.

In this way, the Omnibus model will be suitable for data systems that deal with any kind of sources, including sensors.

2.5.3   Role-Based Model

Role-based model represents a change of focus on how data fusion systems can be modeled and designed. Data fusion systems are specified based on the fusion roles and the relationships among them providing a more fine-grained model for the fusion system. The two members of this generation are the Object-Oriented Model and the Frankel-Bedworth architecture [31]. The role-based model provides a systemic view of data fusion like the JDL model. However, it does not specify fusion tasks or activities. Instead, it provides a set of roles and specifies the relationships among them.

Fig. 2.8 The object-oriented
model for data fusion

1. Object-oriented model: Kokar proposes an object-oriented model for data fusion
   systems. Figure 2.8 is a simplification of the object-oriented model in which four
   roles are identified as:

   - Actor: It is responsible for the interaction with the world, collecting data and
     acting on the environment.
   - Perceiver: After data is gathered, the perceiver assesses such data providing a
     contextualized analysis to the director.
   - Director: The director builds an action plan specifying the system's goals,
     based on the analysis provided by the perceiver.
   - Manager: It controls the actors to execute the plans formulated by the
     director.

2. Frankel-Bedworth architecture: Frankel described an architecture for human
   fusion composed of two self regulatory processes:

   - Local: The local estimation process manages the execution of the current
     activities based on goals and timetables provided by the global process.
   - Global: The global process updates the goals and timetables according to the
     feedback provided by the local process.

   Figure 2.9 shows the Frankel-Bedworth architecture. The local and global
   processes have different objectives and, consequently, different roles.
   The local process tries to achieve the specified goals and maintain the specified
   standards. The local process has the estimator role, which is similar to the
   previous fusion models and includes the following tasks:

   - Sense: Data is gathered by the data sources.
   - Perceive: Stimuli retrieved by sensing are dealt according to its relevance
     (focus), and the Controller is informed which stimuli are being used
     (awareness).

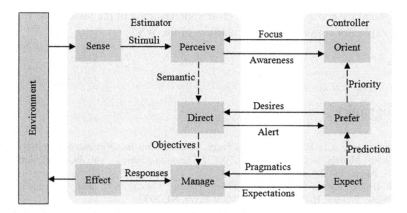

Fig. 2.9 The Frankel-Bedworth architecture

- Direct: Based on the comprehension of the perception (semantics), the Estimator can provide a feedback (alert) to the Controller. The disparity between current situation and desired situation is evaluated. Then, the Estimator is fed forward with desires that specify new goals and timetables.
- Manage: Based on the objectives, the Controller is activated to define what is practical (pragmatics) so the Estimator can provide an appropriate response. Then, the Estimator provides a feedback to the Controller by reporting the expectations about the provided decision (sensitivity).
- Effect selected decisions (responses) are applied and the control loop is closed by sensing the changes in the environment.

Global control process manages the goals of the system during the execution of the local process. The global process has the Controller role; it is responsible for controlling and managing the Estimator role and includes the following tasks:

- Orient: The relevance of sensed stimuli is configured.
- Prefer: Priority is given to the aspects that are most relevant to the goal-achieving behavior, detailing the local goals.
- Expect: Prediction is made and the intentional objective is filtered, determining what is practical to the estimator pragmatics.

The Frankel-Bedworth architecture introduces the notion of a global process separated from the local process. The global control process rules the local process by monitoring its performance and controlling its goals. Moreover, the local process is supposed to perform and implement fusion methods and algorithms to accomplish the system's objectives. This architecture expands the previous models that were concerned only with the local process aspects. In WSN, the global control process will most likely be performed by human beings who feed the network with operation guidelines, whereas the local estimation process should be implemented within the computational system.

Although these models provide a clear understanding of the fusion task, they do not explicitly consider the particularities of the WSN.

Bibliography

1. F. Nakamura, A.F. Loureiro, and C. Frery, "Data Fusion for Wireless Sensor Networks: Methods, Models, and Classifications," *ACM Computing Surveys,* vol. 39, No. 3, Article 9, 2007, August 2007.
2. R.R. Brooks and S. Iyengar, *Multi-Sensor Fusion: Fundamentals and Applications with Software*: Prentice Hall PTR, Upper Saddle River, NJ, 2003.
3. C. Intanagonwiwat, R. Govindan, and D. Estrin, "Directed diffusion: A scalable and robust communication paradigm for sensor networks," in *Proceeding of the 6th ACM Annual International Conference on Mobile Computing and Networking,* Boston, MA, USA, August 2000, pp. 56–67.
4. L. Krishnamachari, D. Estrin, and S. Wicker, "The impact of data aggregation in wireless sensor networks," in *Proceeding of the 22nd International Conference on Distributed Computing Systems Workshops*, Vienna, Austria, July 2002, pp. 575–578.
5. A. Woo, T. Tong, and D. Culler, "Taming the underlying challenges of reliable multihop routing in sensor networks," in *Proceeding of the 1st International Conference on Embedded Network Sensor Systems*, Los Angeles, November 2003, pp. 14–27.
6. "Data fusion lexicon," U. S. D. o. Defence, Ed.: Data Fusion Subpanel of the Joint Directors of Laboratories 1991.
7. J. Llinas and D.L. Hall, "An introduction to multi-sensor data fusion," in *Proceeding of the IEEE International Symposium on Circuits and Systems*, Monterey, CA, USA, May 1998, pp. 537–540.
8. L. Wald, "Some terms of reference in data fusion," *IEEE Transactions on Geoscience and Remote Sensing,* vol. 37, pp. 1190–1193, May 1999.
9. R.C. Luo and M.G. Kay, "Multisensor integration and fusion in intelligent systems," *IEEE Transactions on Systems, Man and Cybernetics,* vol. 19, pp. 901–931, October 1989.
10. K. Kalpakis, K. Dasgupta, and P. Namjoshi, "Efficient algorithms for maximum lifetime data gathering and aggregation" *The International Journal of Computer and Telecommunications Networking,* vol. 42, pp. 697–716, August 2003.
11. A. Boulis, S. Ganeriwal, and M.B. Srivastava, "Aggregation in sensor networks: an energy-accuracy trade-off," in *Proceeding of the 1st IEEE International Workshop on Sensor Network Protocols and Applications,* Anchorage, AK, USA, May 2003, pp. 128–138.
12. A. Hoover and B.D. Olsen, "A real-time occupancy map from multiple video streams," in *Proceeding of the IEEE International Conference on Robotics and Automation*, Detroit, MI, USA, May 1999, pp. 2261–2266.
13. Y.J. Zhao, R. Govindan, and D. Estrin, "Residual energy scan for monitoring sensor networks," in *Proceeding of the IEEE Wireless Communications and Networking Conference*, Orlando, Florida, USA, March 2002, pp. 356–362.
14. X. Luo, M. Dong, and Y. Huang, "On distributed fault-tolerant detection in wireless sensor networks," *IEEE Transactions on Computers,* vol. 55, pp. 58–70, January 2006.
15. J. Polastre, J. Hill, and D. Culler, "Versatile low power media access for wireless sensor networks," in *the 2nd ACM International Conference on Embedded Networked Sensor Systems,* Baltimore, USA, November 2004, pp. 95–107.
16. B. Krishnamachari and S. Iyengar, "Distributed Bayesian algorithms for fault-tolerant event region detection in wireless sensor networks," *IEEE Transactions on Computers,* vol. 53, pp. 241–250, March 2004.

17. B.V. Dasarathy, "Sensor fusion potential exploitation-innovative architectures and illustrative applications," in *Proceedings of the IEEE,* vol. 85, pp. 24–38, January 1997.
18. M.L. Sichitiu and V. Ramadurai, "Localization of wireless sensor networks with a mobile beacon," in *Proceeding of the IEEE International Conference on Mobile Ad-hoc and Sensor Systems*, Pasadena, CA, USA, May 2004, pp. 174–183.
19. A.P. Dempster, "A generalization of Bayesian inference," *Journal of the Royal Statistical Society,* pp. 205–247, March 1968.
20. G. Shafer, *A Mathematical Theory of Evidence* Princeton University Press, 1976.
21. A.J. Pinto, J.M. Stochero, and J.F. Rezende, "Aggregation-aware routing on wireless sensor networks," in *Proceeding of the 9th International Conference on Personal Wireless Communications*, Netherlands, September 2004, pp. 238–247.
22. B. Yu, J. Giampapa, S. Owens, and K. Sycara, "An evidential model of multisensor decision fusion for force aggregation and classification," in *Proceeding of the 8th International Conference on Information Fusion*, Philadelphia, USA. 25–29 July 2005, pp. 8–13.
23. G. Indranil, D. Riordan, and S. Srinivas, "Cluster-head election using fuzzy logic for wireless sensor networks," in *Proceeding of the 3rd Annual Communication Networks and Services Research Conference*, Halifax, Novia Scotia, Canada, May 2005, pp. 255–260.
24. D.S. Friedlander and S. Phoha, "Semantic data fusion for coordinated signal processing in mobile sensor networks," *International Journal of High Performance Computing Applications*, vol. 14, pp. 235–241, April 2002.
25. L. Xiao, S. Boyd, and S. Lall, "A scheme for robust distributed sensor fusion based on average consensus," in *Proceeding of the 4th International Symposium on Information Processing in Sensor Networks*, Los Angeles, California, USA, April 2005, pp. 63–70.
26. L. Xiao, S. Boyd, and S. Lai, "A space-time diffusion scheme for peer-to-peer least-squares estimation," in *Proceeding of the 5th International Conference on Information Processing in Sensor Networks*, Nashville, TN, USA, April 2006, pp. 168–176.
27. Y. Chin-Lung, S. Bagchi, and W. J. Chappell, "Location tracking with directional antennas in wireless sensor networks," in *Proceeding of the IEEE International Microwave Symposium Digest*, Long Beach, CA, USA, June 2005, pp. 4–10.
28. R.E. Kalman, "A new approach to linear filtering and prediction problems," *Journal of Basic Engineering,* vol. 3, pp. 35–45, May 1960.
29. R. Olfati-Saber, "Distributed Kalman Filter with Embedded Consensus Filters," in *Proceeding of the 44th IEEE Conference on Decision and Control*, Seville, Spain, December 2005, pp. 8179–8184.
30. M. Bedworth and J. O'Brien, "The Omnibus model: a new model of data fusion?," *IEEE Aerospace and Electronic Systems Magazine,* pp. 30–36, April 2000.
31. B. Frankel, "Control, estimation and abstraction in fusion architectures: Lessons from human data processing," in *Proceeding of the 3rd International Conference on Data Fusion*, Paris, France, May 2000, pp. 3–10.

Chapter 3
Proposed Centralized Data Fusion Algorithms

Abstract The trend in oil companies nowadays is to integrate the entire well sensors (modern and legacy sensors) with wireless sensor network (WSN). In this work, we introduced a new framework from such sensors using a heterogeneous network of sensors taking in our consideration the WSN's constraints. The framework combined two modules: a Wireless Sensor Data Acquisition (WSDA) module and a Central Data Fusion (CDF) module. A test bed was established from ten acoustic sensors mounted on a closed loop pipeline. The flow rate and the differential pressure were monitored as well. The CDF module was implemented in the gateway using four fusion methods; Fuzzy Art (FA), Maximum Likelihood Estimator (MLE), Moving Average Filter (MAF) and Kalman Filter (KF). The results show that the KF fusion method is the most accurate method. Unlike the other methods, Kalman filter algorithm does not lent itself for easy implementation; this is because it involves many matrix multiplication, division and inversion. Among these 17 matrix operations, there are 10 matrix multiplications, 2 matrix inversions, 4 matrix additions and 1 matrix subtraction. Moreover, these tasks are computationally intensive and strain the energy resources of any single computational node in a WSN. In other words, most sensor nodes do not have the computational resources to complete a central KF task repeatedly. Furthermore, the computational complexity of the centralized KF is not scalable in terms of the network size.

3.1 Introduction

In a centralized fusion system, all data to be combined is sent to a central fusion node that performs the complete fusion task. All data processing is performed in a central unit. In this work, we proposed four centralized fusion algorithms to be implemented in WSN. As a case of study, we propose a remote monitoring framework for sand production in pipelines. Our goal is to introduce a reliable and accurate sand monitoring system.

A. Abdelgawad and M. Bayoumi, *Resource-Aware Data Fusion Algorithms*   37
for Wireless Sensor Networks, Lecture Notes in Electrical Engineering 118,
DOI 10.1007/978-1-4614-1350-9_3, © Springer Science+Business Media, LLC 2012

Produced sand in oil pipelines is a major problem in many production situations since a small amount of sand in the produced fluid can result in significant erosion. In high velocity oil wells erosion is a serious problem since it can erode holes in the pipe in a very short time period [1]. It may cause considerable erosion damage in the well tubing, fittings, separators, valves and other equipments. Produced sand also can result in serious damage to the reservoir, where in some cases the reservoir collapses as a result of the sand production. The most commonly used practice for controlling sand erosion in a gas and oil producing well is simply to limit the production. It can cause poor performance in injection wells, and can lead to lost production. It arises in the case of failure of sand control measures. Sand screening also is a critical part of the mining process [2].

Every year, cleaning and repair operations related to sand production and restricted production rates cost the industry millions of dollars. Quite often producers worry over the consequences of sand production that limits the oil and gas production seriously. Sand usually comes in batches, in other words in large or small quantities. However, the time period for the batches can vary both for the individual well itself as from well to well. When sand has been constantly produced over a period of time one should be aware of the conditions within the reservoir itself. However, a reliable sand monitor system is essential to decide whether sand control measures need to be installed during well operations or not. Sand monitoring allows timely actions to be taken to handle sand production such as:

1. Increasing inspection to detect erosion.
2. Reducing the flow rate or stop the production in some of the extreme cases.
3. Installation of sand handling systems [3, 4].

In order to prevent a high potential incident from occurring, several fields have installed a sand detection system. Installation of a system to monitor and quantify sand production from a well would be valuable to assist in optimizing well productivity and to detect sand as early as possible. Early detection would then lead to possible remedial action that could prevent incidents due to erosion and improve production. By using an efficient monitoring device with a high degree of repeatability and sensitivity, the producers are capable of not only avoiding erosion–corrosion or reservoir damages, but also increasing the oil & gas production. However to be able to do this you need repeatability, sensitivity and also real time measurement. Design to prevent sand erosion is often done by ad-hoc methods that are independent of the sand production rate.

3.2   Sand Measuring in Pipelines

Some sand monitoring devices are located down hole on the production tubing. More commonly, monitoring is undertaken on the topsides pipe work. Two generic types of devices are used to monitor sand production: the intrusive devices and non-intrusive devices.

Fig. 3.1 (a) Intrusive device.
(b) Non-intrusive devices

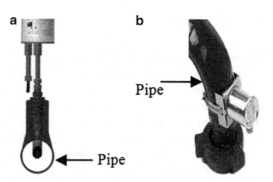

3.2.1 The Intrusive Devices

The intrusive devices can be intrusive sensors, tuning forks or they can be erodible resistance probes as shown in Fig. 3.1a. These are probes inserted through the pipe wall into the flow path. Erodible resistance probes are the most commonly used today among the intrusive types. This type uses the Whetstone Bridge as a principle for measurement techniques, and is a well proven and working principle. However all the intrusive systems have some disadvantages due to their intrusiveness i.e., it is not real time measurement device, as they are not able to give the user a quick respond. Moreover, sand and particles in the flow causes erosion on the probe and it has to be replaced. One type of sand probe is used to detect erosion. Other types are used for continuous monitoring. One type uses an acoustically sensitive crystal to generate an electrical pulse when the inserted portion of the sand probe is struck by a particle of sand. Operation is at ultrasonic frequency to control interference from background flow noise. The energy of the pulse is processed to estimate the amount of sand in the flow stream. The other type is based on measuring the change in electrical resistance of sensing elements, which are eroded by the sand. The measured metal loss is processed to indicate a sand production rate. Sand probes can be periodically polled by a main processor, the data on the quantity of produced sand being transmitted to a data acquisition unit for display and trending.

3.2.2 The Non-intrusive Devices

The non-intrusive devices are clamped onto the pipe wall as shown in Fig. 3.1b. They are acoustic devices that detect the sound of particles impacting the pipe wall. However careful positioning of the device is essential; usually they are installed after a bend. When the flow is passing the bend, particles will be forced out and hit the inside of the pipe wall and generate an ultrasonic pulse. The ultrasonic signal is transmitted through the pipe wall and picked up by the acoustic sensor itself. It may also be necessary to monitor the accumulation of sand within production separators or sand accumulators to determine when water-jetting or flushing needs to be undertaken. Nucleonic level detection devices are typically used.

3.3 Proposed Remote Measuring for Sand in Pipelines

We propose a remote monitoring system for sand in pipelines. Our goal is to introduce a reliable, accurate and low cost sand monitoring system. Figure 3.2 shows the proposed system. The framework combines two modules: a Wireless Sensor Data Acquisition (WSDA) module and a Central Data Fusion (CDF) module. Each of the two modules has a wireless Receiving and Transmission (ReT) module for communication between each others. The framework is designed to collect data from oil pipeline using acoustic sensors (SENACO AS100), Flow Analyzer (MC-II) and Differential Pressure Transmitter (EJA110A) in real time. The data is collected in the gateway i.e., laptop in our case. CDF module is implemented in the gateway using four fusion methods; Fuzzy Art (FA), Maximum Likelihood Estimator (MLE), Moving Average Filter (MAF) and Kalman Filter (KF).

3.3.1 Sensors Used in the Proposed System

1. Acoustic sensor: The Senaco AS100 Sensor monitors high-frequency acoustic emissions. Acoustic emissions travel readily through solid materials such as metal, but are strongly attenuated when traveling through air. As such, the Sensor is immune to airborne interferences and provides a non-invasive method of monitoring process activities [5].The Senaco AS100 Sensor provides an analog output for use. It is primarily used for solids flow detection. However, this device can be used in pump cavitations and fluid leak detection, provided sufficient noise levels are generated. Figure 3.3 shows the AS100 acoustic sensor. Because the AS100 is mounted outside the process, it is completely non-invasive. In hazardous or hygienic environments, this is a great advantage as

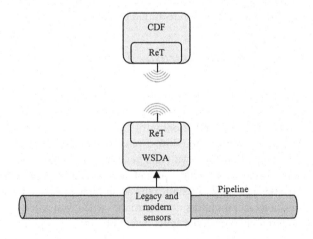

Fig. 3.2 Proposed platform

Fig. 3.3 Senaco AS100
sensor

there is no need for constant cleaning, and concerns about product contamination are eliminated. The Senaco AS100 is also unaffected by abrasive applications.

2. MC-II flow analyzer specifications: The NuFlo Measurement System's Model MC-II Flow Analyzer receives an electronic pulse stream from a turbine flow meter and provides a registration of the totalized flow and an indication of flow rate by utilizing its microprocessor-based circuitry[6]. The totalized flow and the flow rate are displayed on two six-digit Liquid Crystal Display (LCD's). Both displays are properly labeled with respective measurement units. The low current draw of its Complementary Metal–Oxide–Semiconductor (CMOS) microprocessor-based circuitry permits MC-II to run for 3–5 years on single battery. MC-II has the advantage of being battery powered and enclosed in non-corrosive weatherproof housing, deemed ideal for use in remote locations. Figure 3.4 shows the MC-II Flow Analyzer.

3. EJA110A differential pressure transmitter: The accurate measurement of Differential Pressure (DP) is required at many points in the oilfield. In general, DP is defined as a measurement of fluid force subtracted from a higher measurement of fluid force (in terms of pounds per square inch).Yokogawa Electric Cooperation Model EJA110A differential pressure features high performance, durability, and reliability. The pressure detector, the core of the transmitter, uses a silicon resonant sensor that has proven to be highly reliable in the field and offers a complete product lineup. Figure 3.5 shows the EJA110A differential pressure transmitter. The EJA series uses a silicon resonant sensor formed from monocrystal silicon, a perfect material which has no hysterics in pressure or temperature changes. The sensor minimizes overpressure, temperature change, and static pressure effects, and thus offers unmatched long-term stability. EJA110A is a compact and light-weight design. It has also field bus communication capability. Field bus is a digital two-way communication system. It is a revolutionary technology for configuring instrumentation control systems and a promising successor to the standard 4–20 mA analog communications used in most field instruments today [7].

Fig. 3.4 The MC-II flow
analyzer

Fig. 3.5 EJA110A
differential pressure

3.3.2 WSDA Framework

Our proposed WSDA framework includes three components: a signal Conditioning
and Digitizing (CoD) module, a wireless Receiving and Transmission (ReT)
module, and a Management and Control (MaC) module. While different
implementations of the ReT module are possible, this paper discusses in detail an
implementation based on TinyOS [8] and Crossbow MICA2 motes [9]. The CoD
module has been implemented to work with three legacy sensors: acoustic sensor,
flow meter and differential pressure transmitter. These devices produce different
types of analog signals: the acoustic sensor and the flow meter generate Direct
current (DC) voltages while the differential pressure generates DC current outputs

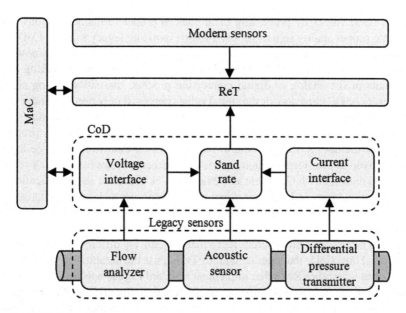

Fig. 3.6 WSDA framework

proportional to the measured differential pressure. Figure 3.6 shows the proposed WSDA. The CoD module permits: (a) various legacy sensors to provide their data readings in a unified way to the ReT module for transmission. (b) Received data/control signals to be converted back to their respective analog signal forms and ranges (in voltages or DC currents) native to those legacy sensors involved. Upon reception, it interfaces with the CoD module, which converts the received values before forwarding to corresponding legacy sensors. In addition, it conditions and digitizes the analogous sensor readings to the given resolution for transmission. The MaC module serves to manage and control, in a unified manner, the sensor network formed and the sensor data acquisition rates. The proposed WSDA framework is readily applicable to various engineering applications where legacy and modern sensors are co-exist.

As it is responsible for wireless signal reception and transmission, the ReT module may be implemented in various fashions. For example, one may build it using a Crossbow's mote which is governed by the XMesh communication protocol implemented in TinyOS. Alternatively, one may choose to implement it by means of a different RF gear set following Zigbee. Another option is to employ a pair of SAW-based RF transmitter and receivers with an integrated encoder and decoder logics, suitable for remote control/command, security, and automation.

The CoD module provides a linkage between legacy sensors and the ReT module. It includes multiple components: a hardware circuit that conditions and normalizes analogous signals from legacy sensors, a logic unit to convert conditioned signals to digital forms, and software which processes digitized values into ones ready for transmission by ReT. The same components are charged to deal with received values by processing them back to proper forms before converting them into proper analog signals for use by corresponding legacy sensors. CoD does not need to be aware of the details of the radio transmission but need only know how to pass the data to/from ReT on transmission. Similarly, ReT knows nothing about the details of the analog to digital conversion process, the native analog signals (in voltages or DC currents), or the signal values ranges. It only needs to know that a CoD has registered with it and is to provide the data in a normalized format.

As management and control of sensor data are basic to WSDA, the proposed framework includes a MaC module to facilitate sensor data readings to be logged and displayed in a uniform fashion, no matter whether they are from legacy sensors or modern ones. The MaC module also enables quick additions and modifications to data collection types and amounts when involved sensors and devices change.

The WSDA framework allows a level of interchangeability between components. The hardware circuits used to condition and normalize a 4–20 mA signal can be reused with different ReTs that use entirely different communication protocols. Thus, CoDs that operate with ReTs that use the default TinyOS protocols can be used with other ReTs that implement standard Zigbee application protocols. While certain modern sensors incorporate the functionality of CoD, this framework integrates data acquisition over those modern sensors together with their legacy counterparts through ReT with acquired data managed and controlled MaC. As can be seen in Fig. 3.6, the proposed framework is suitable for any engineering application where various types of sensors co-exist because its ReT interfaces seamlessly with legacy sensors via CoD and with modern sensors directly.

1. ReT module: As a proof-of-concept attempt, one implementation of the ReT module has been accomplished by using Crossbow Technologies MICA2 motes and the TinyOS operating system. As the hardware component of our ReT implementation, MICA2 motes employ the Chipcon CC1000 FSK modulated radio and come in three models, according to their RF frequency bands: the MPR400 (915 MHz), MPR410 (433 MHz), and MPR420 (315 MHz). All models utilize an Atmega128L micro-controller and a frequency-tunable radio with an extended range. A variety of modern sensors and data acquisition boards can be connected to a MICA2 mote. Communication is two-way between the ReT and connected CoD. The ReT provides an interface that can be used to pass commands to the ReT and any connected sensors. For example, radio transmission power can be adjusted by sending a command to the ReT. The interface between the ReT and other components in the framework reduces the effort required to link with existing industrial control systems. The application programmer needs only to implement the ReT to CoD interface in order to link a new device into the ReT's network. In a similar fashion, it is possible to replace an existing ReT with a ReT that communicates to an entirely different network. Software reusability has long been sought for improved productivity and lowered development and testing costs. It can be enhanced through a set of proven solutions for recurrent problems. In general, software design patterns do not specify implementation details, making it possible to accommodate various

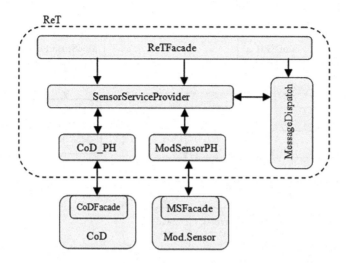

Fig. 3.7 ReT component design

scenarios via reusing their high-level abstractions [10]. ReT is implemented to consist of three software design patterns: the Facade, ServiceInstance, and Dispatcher patterns [11], as illustrated in Fig. 3.7. The Facade design pattern is used when one needs a single, unified interface to a collection of sub-services. In our case, the entry point to ReT for other applications is the ReTFacade application interface.

Applications use this interface to post requests for launching data acquisition from the devices, initiating data transmission/reception, and registering new devices. The ReTFacade component passes these messages either to the SensorServiceProvider for administrative requests or to the MessageDispatcher for all other messages. The SensorServiceProvider is responsible for keeping track of each instance of the CoD or ModernSensor acquiring data. The MessageDispatcher is aware of each sensor connected to the system via a registration interface in the dispatcher which is called when SensorService-Provider instantiates a new sensor instance. Messages are then routed directly to the sensor by the MessageDispatcher component. ReT is aware only of a Proxy object for each sensor. This instance of the Proxy pattern handles the interface between ReT and the devices. A single proxy will be in place for CoD while multiple instances may exist for each ModernSensor to which the ReT communicates.

2. CoD module: CoD may collect data from more than one instance of either a Voltage Signals device or Current Signals component, as depicted in Fig. 3.8. Here, a variation of the ServiceInstance pattern is developed, with CoDFacade also serving as the ServiceProvider component of a multi-to-multi resource multiplexer/demultiplexer. The ServiceProvider component realizes the service type of corresponding sensors, and it varies with the type of sensors. As legacy sensors and measurement devices usually produce analog outputs in the form of

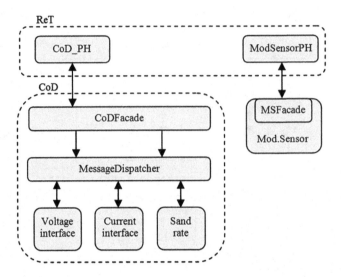

Fig. 3.8 CoD component design

voltage or direct current (4–20 mA) signals, our CoD module includes circuits for output conditioning and amplification to a specified voltage range before ADC conversion is applied [12]. Three sets of circuit design have been devised and implemented: one for voltage conditioning and amplification, another for direct current signal conditioning and conversion to voltage, and one to calculate the sand production rate (i.e., sand rate module) as follows.

- Voltage outputs: An amplifier circuit is designed to amplify the flow analyzer's output from 1 to 2.5 V before digitized. As illustrated in Fig. 3.9, an LMC6484 CMOS quad rail-to-rail input and output op-amp is used in the amplifier circuit, providing a common-mode range that extends to both supply rails [13]. This rail-to-rail performance, combined with excellent accuracy, makes it unique among rail-to-rail input amplifiers. It is ideal for systems, such as data acquisition, that require a large input signal range. Maximum dynamic signal range is assured in low voltage and single supply systems by the LMC6484's rail-to-rail output swing, which is guaranteed for loads down to 600 Ω. This guaranteed low load characteristic and its low power dissipation make LMC6484 especially well-suited for battery-operated systems.

  Given the gain of the circuit is calculated by the next equation, the values of resistors R1 and R2 equal to 10 and 15 KΩ respectively. The circuit brings the input voltage V_{in} of 1 V to the output voltage V_{out} of 2.5 V, as required:

$$\frac{V_{out}}{V_{in}} = 1 + \frac{R_2}{R_1} \qquad (3.1)$$

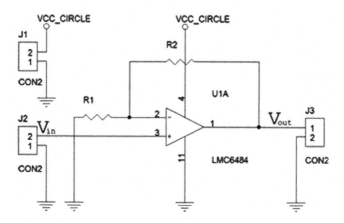

Fig. 3.9 Voltage amplification diagram

Fig. 3.10 Voltage divider
diagram

It should be noted that this designed circuit is universal for conditioning and
amplifying voltage outputs of any legacy sensor, no matter what its output
voltage value might be. As the ADC converter takes 2.5 V as its input,
a legacy sensor's voltage output can always be rectified to the proper range
before conversion through choosing appropriate R1 and R2 [14].

- Voltage inputs: For a legacy sensor with voltage inputs to receive control
  information or data wirelessly, CoD employs a voltage dividing circuit shown
  in Fig. 3.10, with R1 and R2 governed by the equation below:

$$V_{out} = V_{in} \frac{R2}{R1 + R2} \tag{3.2}$$

The received information and data, after being converted to an analogous
form, can be conditioned to match the voltage range suitable for the sensor
input.

- Current outputs: The 4–20 mA current loop has been a popular sensor output
  form for industrial and process sectors alike. Its popularity comes from
  its ease of use, its performance, and its simple wiring. Both the supply voltage
  and the measuring current are over the same two wires. This current loop
  makes cable break detection simple: if the current drops to zero, a cable break
  happens. Additionally, the current signal is immune to any stray electrical
  interference, deemed particularly important to sensors applied to harsh

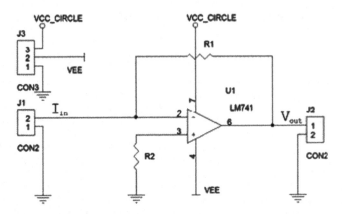

Fig. 3.11 Current-to-voltage converting circuit

environments. Those advantages make the current loop output common to many legacy sensors.

A current-to-voltage converter translates current signals to proportional voltage outputs. It includes an operational amplifier for simple linear signal processing and a resistor for dissipating current, as depicted in Fig. 3.11. The resistance between the operational amplifier's input and output determines the voltage range for specific current signals. In current-to-voltage converters that handle a range of currents, design consideration accounts for the DC offset caused by both the input device and the operational amplifier. The output voltage of the circuit is calculated by the equation of $V_{out} = R1 \times I_{in}$, giving rise to Vout in the range of 0.4–2 V, for R1 $= 100\ \Omega$ when the current loop of 4–20 mA is applied. To minimize the bias current, R2 is chosen to equal R1. This circuit yields a desirable voltage range for ADC conversion, according to the current loop reading produced by a legacy sensor. The circuit shown in Fig. 3.11 is base on the LM741 op-amplifier [15]. Since the 4–20 mA current loop is the most popular output form of sensors deployed in many engineering applications, this CoD circuit is generally suitable for those applications.

- Current inputs: For a legacy sensor with current loops as its inputs, a companion circuit to Fig. 3.12 is needed for converting voltages which are obtained from digitalized data or control information received wirelessly to 4–20 mA DC currents. A voltage-to-current converting circuit can be included in the CoD module. The circuit was an application example given in the AM422 op-amplifier data sheet [16]. The AM422 is a low cost monolithic voltage–to–current converter specially designed for analog signal transmission. In order to get an output current (Iout) range of 4–20 mA with the input DC voltage (Vin) ranging from 0.4 to2.0 V, the circuit discrete components of Fig. 3.12 are specified as follows: R0 $= 25\ \Omega$, R3 $=$ R4 $= 33$ KΩ, RSET $= 2.64$ KΩ, R5 $= 40\ \Omega$, R1/R2 $= 2.25$, RL $= 0$–$500\ \Omega$, C1 $= 2.2\ \mu$F.

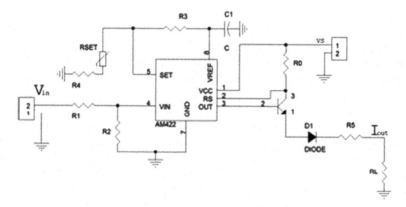

Fig. 3.12 Voltage-to-current converting circuit

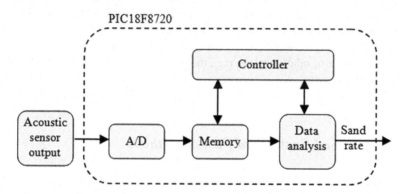

Fig. 3.13 Sand rate module

- Sand rate module: The output of the acoustic sensor is used to calculate the sand production rate using a PIC18F8720 microcontroller [17]. Data is analyzed using descriptive statistic algorithm. Descriptive statistic is used to describe or summarize the data. Descriptive statistic is used throughout data analysis in a number of different ways. Simply stated, they refer to Minimum, Maximum, Mean, and Standard Deviation and numbers of valid cases of one variable. Descriptive statistic is important in data cleaning. It is regularly used during analysis to keep an eye on the variables being used, especially when a considerable number are being studied. Descriptive analysis can often be presented more accurately for the continuous variables than for categorical variables because of lost information from collapsing it into categories. Figure 3.13 shows the block diagram of the sand rate module.

3. MaC module: Data values collected from sensors and measurement devices are transmitted over a WSDA system normally to one or multiple designated sink

node. To facilitate management and control on data acquisition, a generic management portal is developed as part of our WSDA framework. The MaC module enables quick additions and modifications to data collection types and amounts when involved sensors and devices change. It interfaces with an application server, called XSERVE, which is responsible for collected data logging into an archival database, to permit data displays and processing. MaC consists of three units: (a) The well data sensor portlet, (b) The charts portlet, and (c) the sensor network configuration portlet that together realize required functions to manage WSDA. The three portlets are constructed using Gridsphere, which is an open-source Java-based web portal [18], and its accompanying GridPortlets. Gridsphere eases the development and deployment of portlet web applications for efficient administration. It provides the tools needed for fast development and prototyping of the MaC portlets. The sensor network configuration portlet enables one to modify the communication parameters governing connections to XSERVE and the database server. For example, the user can select the server Internet Protocol (IP) address, Transmission Control Protocol (TCP) port used to connect to the database, and the TCP port used to connect to the sensors. It can also be used to specify the appearance of other portlets; e.g., the number of raw data packets shown in the Xsensor porlet.

3.3.3 Proposed Centralized Fusion Methods

In our proposed platform, we have ten acoustic sensors distributed on the pipeline. Each sensor has a sand rate interface to calculate the sand production rate, the flow rate and differential pressure are monitored as well. At the gateway, we are receiving ten different values from each sensor. In order to increase the associated confidence, we need to fuse all the data coming from the sensors. In this work we propose four different fusion algorithms: Fuzzy Art (FA), Maximum Likelihood Estimator (MLE), Moving Average Filter (MAF) and Kalman Filter (KF).

1. Fuzzy art: FuzzyART data fusion technique [19] employs FuzzyART neural network modules to fuse measurements into a coherent estimate. FuzzyART modules are a class of unsupervised neural network systems where training is performed without the presence of a teacher. Moreover, training is done online, which makes the system highly adaptive to changes in the input vector. Neural network modules fit in typical sensor network systems since it shares similar characteristics, such as, distributed processing and data storage, adaption to changes in environment, and resilience to noise and corrupted data. After the input vector is presented to the FuzzyART fusion center, it will cluster the information into categories. The weight vector of the trained network has a geometric meaning where each category is bounded by a rectangle that contains all the inputs that are close to each other. FuzzyART fuses

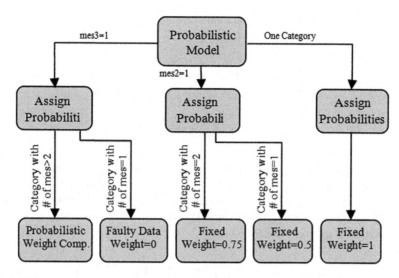

Fig. 3.14 FuzzyART decision tree

measurements by assigning a probabilistic weight according to two metrics, spatial correlation and consensus vote. That is, measurements that are geometrically close to each other should be given high probabilistic weight and the more estimates we have in a group the more confident we are with this group. Furthermore, FuzzyART detect erroneous measurements and assign them a zero weight to prohibit them from contaminating the estimation process. To automate this process probabilistic weight is assigned according to a decision tree shown in Fig. 3.14.

If all measurements reside in a single category then a fixed probabilistic of one will be assigned to this category which is divided evenly among the candidates of this category. If measurements are divided into two categories where one category contains two measurements and the other contains only one measurement, then we cannot rule out the fact that the category with one measurement contains an erroneous measurement. In this case, the category with two measurements is assigned a weight of 0.75 which is divided evenly between the two measurements. The category with one estimate is assigned a weight of 0.25. If a category or more contains more than two measurements and if zero or more categories contain one measurement, then we definitely can assume that categories with one measurement are faulty and should be assigned a zero weight. On the other hand, categories with two measurements or more will be assigned a probabilistic weight following the methodology listed below. FuzzyART assigns probabilistic weights according to spatial correlation and consensus vote. For case number 3 the probabilistic weight assignment is performed according to the following methodology: The higher number of measurements belonging to the same category (consensus vote) the higher the weight should be (Eq. 3.3). Moreover, lower standard deviation between the measurements (spatial correlation) belonging to the same group should be given also a higher probability (eq).

$$P_m = \frac{n_{mess_{cat}}}{tot_{n_{inputs}}} \qquad (3.3)$$

$$P_s = 1 - \frac{STD_{cat}}{tot_{STD}} \qquad (3.4)$$

where n_mess_cat is the number of measurements in a category and tot_n_input is the total number of inputs, and STD_cat is the standard deviation of the measurements in the same category and tot_STD is the total standard deviation of all categories. Furthermore since P_a and P_m are independent their joint probability is given by the following:

$$P(s,m) = P_m \times P_s = \left(\frac{n_{mess_{cat}}}{tot_{n_{inputs}}}\right) \times \left(1 - \frac{STD_{cat}}{tot_{STD}}\right) \qquad (3.5)$$

The final weight given to each category is the normalization of Eq. 3.5 given by (3.6) where n_comm is the number of committed categories:

$$P_j = \frac{P_j(m,s)}{\sum_i^{n_{comm}} P_i(m,s)} \qquad (3.6)$$

The final estimate is given by multiplying each measurement by its corresponding weight and summing them all up as given by Eq. 3.7.

$$\tilde{x} = \sum_{i=0}^{N} x(i) \times P_i \qquad (3.7)$$

2. Maximum likelihood estimator (MLE): Maximum Likelihood Estimator is a method for choosing estimator of parameters that avoids using prior distributions and loss functions. It is also a statistical estimator that can be used to estimate a model's unknown parameter values from data [20]. Suppose that x1, ..., xn are the observed data coming from n sensors. And suppose that X1, ..., Xn are following a random sample from a normal distribution with unknown mean μ and variance σ^2. The parameter set is then $\theta = (\mu, \sigma^2)$. For all observed values x1, ..., xn, the likelihood function

$$f_n\left(x|\mu, \sigma^2\right) = 1/(2\pi\sigma^2)^{(n/2)} \quad exp[-1/(2\sigma^2) \; \Sigma_{(i-1)}^n (X_i - \mu)^2] \qquad (3.8)$$

This function must now be maximized over all possible values of μ and σ^2.

Where $-\infty < \mu < \infty$ *and* $\sigma^2 > 0$. Instead of maximizing the likelihood function $f_n(x|\mu, \sigma^2)$ directly, it is easier to maximized $\log f_n(x|\mu, \sigma^2)$.

$$L(\theta) = \log f_n(x|\mu, \sigma^2) = -\frac{n}{2} \log (2\pi) - \frac{n}{2} \log \sigma^2 - \frac{1}{2\sigma^2} \sum_{i=1}^{n} (x_i - \mu)^2 \quad (3.9)$$

Therefore, in principle, it is possible to derive the sampling distribution of each estimator of θ. For example, if $X1, \ldots, Xn$ form a random sample from a normal distribution with mean μ, and variance σ^2, then it is known that the sample mean $\bar{X}n$ is the MLE of μ. Furthermore, it was found that the distribution of $\bar{X}n$ is a normal distribution with mean μ, and variance σ^2/n [21].

3. Moving average filter (MAF): The Moving Average Filter (MAF) is broadly implemented in fusion. It is optimal for reducing random white noise at the same time as retaining a sharp step response. This filter computes the arithmetic mean of a number of input measurements to produce each point of the output signal. A slight improvement in computational efficiency can be achieved if we perform the calculation of the mean in a recursive fashion. A recursive solution is one which depends on a previously calculated value. To illustrate this, consider the following development: Suppose that at any instant k, the average of the latest n samples of a data sequence, Xi, is given by:

$$\bar{X}_k = \frac{1}{n} \sum_{i=k-n+1}^{k} X_i \quad (3.10)$$

Similarly, at the previous time instant, k−1, the average of the latest n samples is:

$$\bar{X}_k = \frac{1}{n} \sum_{i=k-n}^{k-1} X_i \quad (3.11)$$

Therefore,

$$\bar{X}_k - \bar{X}_{k-1} = \frac{1}{n} \left(\sum_{i=k-n+1}^{k} X_i - \sum_{i=k-n}^{k-1} X_i \right)$$

$$= \frac{1}{n} (X_k - X_{k-n}) \quad (3.12)$$

This on rearrangement gives:

$$\bar{X}_k = \bar{X}_{k-1} + \frac{1}{n} (X_k - X_{k-n}) \quad (3.13)$$

Fig. 3.15 Moving window of
n data

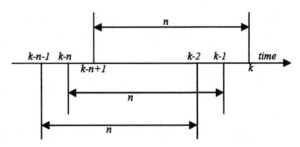

This is known as a moving average because the average at each kth instant is
based on the most recent set of n values. At any time, a moving window of
n values is used to calculate the average of the data sequence as shown in
Fig. 3.15.

4. Kalman filter (KF): The Kalman filter is a set of mathematical equations that
   provides an efficient computational solution to discrete time data filtering
   problems, in essence removing extraneous noise from a given stream of data.
   The filter is very powerful in several aspects: it supports estimations of past,
   present, and even future states, it is an optimal estimator in the case of Gaussian
   uncertainties, and it can do so even when the precise nature of the modeled
   system is unknown. Moreover, the Kalman filter is the best linear estimator for
   any other distributions. Kalman filter is based on linear dynamical systems
   discredited in the time domain. It is modeled on a Markov chain built on linear
   operators perturbed by Gaussian noise. The state of the system is represented as a
   vector of real numbers.

3.4 Simulation and Experimental Results

In this section we provide the output performance of the proposed platform. A test
bed was established using ten acoustic sensors to collect the data from the same
pipe. Figure 3.16 shows the test bed Platform. In order to get the accurate sand
production rate, the flow rate and differential pressure should be monitored as well.
The global flow rate is measured using one MC-II flow analyzer sensor and the
global differential pressure is measured using one differential pressure sensor. Both
sensors are mounted on the test bed. In the experimental tests, the sand is injected in
the test bed by a certain rate using an injector with a known flow rate and pressure.
Each sand rate module calculates the sand rate. The sand rates, the global flow rate
and the global pressure are digitized for wireless transmission using the WSDA
module. The data is collected in the gateway i.e., laptop in our case. CDF module at
the gateway is used to fuse all the ten sand rates in order to improve the system
output. The proposed system has been validated experimentally using different sand
rate, flow rate and pressure.

Fig. 3.16 The testbed platform

Table 3.1 Average percentage error for scenario I

Sensor #	Sensor 1	Sensor 2	Sensor 3	Sensor 4	Sensor 5
Avg.% error	3%	19%	12%	7%	9%
Sensor #	Sensor 6	Sensor 7	Sensor 8	Sensor 9	Sensor 10
Avg.% error	15%	12%	6%	11%	5%

Table 3.2 Average percentage error for scenario II

Sensor #	Sensor 1	Sensor 2	Sensor 3	Sensor 4	Sensor 5
Avg.% error	21%	12%	6%	11%	3%
Sensor #	Sensor 6	Sensor 7	Sensor 8	Sensor 9	Sensor 10
Avg.% error	5%	4%	15%	17%	10%

In the experimental results, each result is the average of ten runs. Table 3.1 (Scenario I) shows the average percentage error between the observed sand rate, from each sand rate module, and the actual sand rate. The sand was injected with 20 G/s (Gram/second), flow rate 25 G/min (Gallon/Minute) and pressure 400 psi (pound per square inch).

(Scenario II) shows error for another scenario: The sand was injected with 15 G/s (Gram/second), flow rate 30 G/min (Gallon/Minute) and pressure 400 psi (pound per square inch).

Tables 3.1 and 3.2 show that the error for each individual sand rate module is very high and inconsistent with others. That is leading us to implement fusion to increase the associated confidence [19, 21].

Tables 3.3 and 3.4 show average percentage error, maximum percentage error and the standard deviation of four different fusion methods (FA, MLE, MAF and KF).

Table 3.3 Comparison
between fusion methods for
scenario I

Method	Avg.% error	Max% error	Stdev.
FA	4.06	9.67	3.02
MLE	6.24	11.06	3.94
MAF	5.08	12.54	4.06
KF	1.17	3.34	1.01

Table 3.4 Comparison
between methods for scenario II

Method	Avg.% error	Max% error	Stdev.
FA	4.01	10.77	3.01
MLE	6.33	8.87	3.88
MAF	5.14	9.13	3.97
KF	1.03	3.11	1.12

The results show that the average percentage error is decreased by using the fusion methods for scenario I and II. The results show that the KF fusion method is the more accurate method. However the FA, MLE and MAF methods are less complex than the KF to be implemented in WSN. MLE and MAF need accumulation operation only. FA needs ADD, OR, DIVIDE and COMPARE operations [22, 23]. Unlike the other methods, Kalman filter algorithm does not lend itself for easy implementation; this is because it involves many matrix multiplication, division and inversion. Altogether computations of an estimate involve 17 matrix operations. Among these 17 matrix operations, there are 10 matrix multiplications, 2 matrix inversions, 4 matrix additions and 1 matrix subtraction. Moreover, these tasks are computationally intensive and strain the energy resources of any single computational node in a WSN. In other words, most sensor nodes do not have the computational resources to complete a central KF task repeatedly. Next chapters we will propose a low power Distributed Kalman Filter (DKF).

Bibliography

1. I. A. Allahar, "Acoustic Signal Analysis for Sand Detection in Wells with Changing Fluid Profiles," in *Society of Petroleum Engineers*, pp. 103–111, October 2003.
2. J. Sheldon, R. Kube, and Z. Hong, "Oil sand screen modelling using partial least squares regression," in *Proceeding of the IEEE International Conference on Automation and Logistics*, Oingdao, China, September 2008, pp. 2936–2940.
3. A. I. Shamma'a, R. T. A. Shaw, and J. Lucas, "On line EM wave sand monitoring sensor for oil industry," in *Proceeding of the 33rd European Microwave Conference*, Munich, Germany, October 2003, pp. 535–538.
4. A. Huser and O. Kvernvold, "Prediction of Sand Erosion in Process and Pipe Components," in *Proceeding of the 1st North American Conference on Multiphase Technology*, Banff, Canada, August 1998, pp. 134–139.
5. Milltronics Inc., "Senaco AS100 Acoustic Sensor," http://www.lesman.com/unleashd/catalog/belt/0460-en-00.pdf.
6. NuFlo Measurement Systems, "User manual," http://www.cam.com/content/products.

7. Yokogawa Electronic Corporation, "EJA110A Differential Pressure Transmitter," http://www.yokogawa.com/fld/PRESSURE/EJA/fld-eja110a-01en.htm.
8. P. Levis and S. Madden, "TinyOS: An operating system for wireless sensor networks," in *Ambient Intelligence Conference*, Eindhoven, Netherlands, November 2004, pp. 123–129.
9. Crossbow Technology, "MICA2 Datasheet," http://www.xbow.com.
10. F. Rincon, F. Moya, and J. Barbra, "Model Reuse through Hardware Design Patterns," in *Proceeding of the IEEE Design, Automation, and Test in Europe Conference and Exhibition*, Munich, Germany, March 2005, pp. 324–329.
11. D. Gay, P. Levis, and D. Culler, "Software Design Patterns for TinyOS," in *ACM Transactions on Embedded Computing Systems*, May 2007.
12. J. Hauer, P. Levis, V. Handziski, and D. Gay, "TinyOS Extension Proposal 101: Analog-to-Digital Convertors (ADCs)," http://www.tinyos.net/tinyos-2.x/doc/pdf/tep101.pdf.
13. National Semiconductor Corporation, "LMC6484 CMOS Quad Rail-to-Rail Input and Output Operational Amplifier," http://www.datasheetcatalog.com/.
14. A. Abdelgawad, A. Lewis, M. Elgamel, F. Issa, N. F. Tzeng, and M. Bayoumi, "Remote Measuring of Flow Meters for Petroleum Engineering and Other Industrial Applications," in *International Workshop on Computer Architecture for Machine Perception and Sensing*, Montreal, Canada, March 2007, pp. 99–103.
15. National Semiconductor Corporation, "LM741 Operational Amplifier," http://www.national.com/.
16. Analog Microelectronics, "Vlotage/Current Converter AM422," http://www.analogmicro.de/.
17. Microchip Technology, "PIC18F8720 datasheet," http://ww1.micr-ochip.com/downloads/en/devicedoc/39609b.pdf.
18. GridSphere, http://www.gridsphere.org.
19. A. Abdelgawad, Z. Merhi, M. Elgamel, and M. Bayoumi, "Multisensor data fusion methods for petroleum engineering applications," in *Proceeding of the IEEE Sensors Applications Symposium*, New Orleans, Louisiana, USA, March 2009, pp. 265–268.
20. M. H. DeGroot and M. J. Schervish, *Probability and Statistics*: Addison Wesley 2002.
21. A. Abdelgawad, Z. Merhi, M. Elgamel, M. Bayoumi, and A. Zaki, "Data fusion framework for sand detection in pipelines," in *Proceeding of the IEEE International Symposium on Circuits and Systems*, Tibia, Taiwan, May 2009, pp. 2173–2176.
22. A. Abdelgawad and M. Bayoumi, "Sand Monitoring in Pipelines Using Distributed Data Fusion Algorithm," *IEEE Sensors Applications Symposium, SAS 2011*, 22–24 Feb. 2011.
23. A. Abdelgawad and M. Bayoumi, "Remote Measuring for Sand in Pipelines Using Wireless Sensor Network," *IEEE Transactions on Instrumentation and Measurement*, vol.60, no.4, pp.1443–1452, April 2011.

Chapter 4
Kalman Filter

Abstract This chapter has briefly discussed the need of the DKF and introduced the literature work of the DKF. Most DKF methods proposed in the literature rely on consensus filters algorithm. The convergence rate of such distributed consensus algorithms typically depends on the network topology and the weights given to the edges between neighboring sensors. The next chapter proposes a low power DKF. The proposed DKF is based on a fast polynomial filter to accelerate distributed average consensus in static network topologies. The idea is to apply a polynomial filter on the network matrix that will shape its spectrum in order to increase the convergence rate by minimizing its second largest eigenvalue. Fast convergence can contribute to significant energy saving.

In 1960, R.E. Kalman published his famous paper presenting a recursive solution to the discrete-data linear filtering problem [1]. The Kalman filter, since that time, has been the subject of extensive research and application, particularly in the area of autonomous or assisted navigation. The Kalman filter is a set of mathematical equations that provides an efficient computational solution to discrete time data filtering problems, in essence removing extraneous noise from a given stream of data. The filter is very powerful in several aspects: it supports estimations of past, present, and even future states, it is an optimal estimator in the case of Gaussian uncertainties, and it can do so even when the precise nature of the modeled system is unknown. Moreover, the Kalman filter is the best linear estimator for any other distributions. Kalman filter is based on linear dynamical systems discredited in the time domain. It is modeled on a Markov chain built on linear operators perturbed by Gaussian noise. The state of the system is represented as a vector of real numbers. At each discrete time increment, a linear operator is applied to the state to generate the new state, with some noise mixed in, and optionally some information from the controls on the system if they are known. Then, another linear operator mixed with more noise generates the visible outputs from the hidden state. The Kalman filter may be regarded as analogous to the hidden Markov model, with the key difference that the hidden state variables are continuous, as opposed to being discrete in the hidden Markov model.

A. Abdelgawad and M. Bayoumi, *Resource-Aware Data Fusion Algorithms*
for Wireless Sensor Networks, Lecture Notes in Electrical Engineering 118,
DOI 10.1007/978-1-4614-1350-9_4, © Springer Science+Business Media, LLC 2012

Additionally, the hidden Markov model can represent an arbitrary distribution for the next value of the state variables, in contrast to the Gaussian noise model that is used for the Kalman filter. There is a strong duality between the equations of the Kalman Filter and those of the hidden Markov model.

Kalman filters are basically calcified as Central Kalman Filters (CKF) and decentralized Kalman filters. A Kalman filter is said to be a central Kalman filter, if all measurements are processed by this Kalman filter. A decentralized Kalman filter is defined as a kind of Kalman filter, in which the measurements are first processed through different local Kalman filter. Their estimations are sending into a fusion center data fusion. If the measurements come from different kind of sensors, it is favorable to use a decentralized Kalman filter to estimate the system state. The decentralized Kalman filter may not only reduce the calculation complexity but also improve the estimate accuracy. A Decentralized Kalman filter does not require any form of central processing facility or centralized communications medium. Each sensing node implements its own local Kalman filter to arrive at a partial decision which it then broadcasts to every other node. Each node then assimilates this received information to arrive at its own local estimate of the system state. The problem of decentralized Kalman filtering was first solved by Speyer [2] in 1979. It was independently resolved by Rao, Durrant-Whyte, and Rao in [3]. A Decentralized Kalman filter requires a complete network with all-to-all links. This solution is not scalable for large-scale sensor networks due to its $O(n^2)$ communication complexity (n is the number of sensors/nodes). Whenever, WSN has finite battery lifetime and thus limited computing and communication capabilities, so decentralized Kalman filter is not applicable in WSN. In particular, it is preferred to avoid decentralized Kalman filter which requires communication between all nodes. Thus, Distributer Kalman Filter (DKF) is preferred where only communication with neighboring nodes is required. By adopting such communication structures we avoid the "cocktailparty" effect, as we can establish spatial clusters in which just a few communication links need to be established. Not only better scalability properties are achieved from the communication side, but also from a computational point of view, as the formation of such clusters also establish a natural hierarchy of computation of estimates in general. In DKF each node only talks to its neighbors, under the assumption that each node has either $O(log(n))$ or $O(1)$ neighbors, the communication cost of this class of DKF is $O(n\ log(n))$, or $O(n)$ which are both scalable in n. Thus, distributed algorithm is the most suitable algorithm to implement Kalman filter in wireless sensor networks.

4.1   Wireless Sensor Network Representation

In wireless sensor network, there is a link between two nodes when packets can be successfully delivered from one node to the other. A wireless sensor network is called connected if for two arbitrary nodes, there is a route, which consists of such links, from one to the other. Traditional work on connectivity analysis of wireless sensor networks often focuses on finding a critical transmission range to keep the network connected. However, some low-cost sensor nodes may not support power-adaptive transmissions. On the other hand, changing the transmission range can be reformulated as changing the density of the sensor networks, in which each node is using fixed transmitted power [4]. Recently, random graph theory is introduced into the modeling of sensor networks with uncertain features. A random graph often can be imagined as a living organism which evolves with time. By giving a set of vertices in advance, the edges are generated according to some randomization rules [5].

4.2 Introduction to Graph Theory

Graph theory can be said to have its beginning in 1736 when EULER considered the Königsberg bridge problem (see Fig. 4.1): Is there a walking route that crosses each of the seven bridges of Königsberg exactly once? It took 200 years before the first book on graph theory was written. This was done by KÖNIG in 1936. Since then graph theory has developed into an extensive and popular branch of mathematics, which has been applied to many problems in mathematics, computer science, and other scientific and nonscientific areas. There seem to be no standard notations or even definitions for graph theoretical objects. This is natural, because the names one uses for these objects reflect the applications. So, for instance, if we consider a communications network as a graph, then the computers, which take part in this network, are called nodes rather than vertices or points.

Graph theory has abundant examples of *NP*-complete problems. Intuitively, a problem is in *P* (Solvable- by an algorithm- in polynomial many steps on the size of the problem instances) if there is an efficient (practical) algorithm to find a solution to it. On the other hand, a problem is in *NP* (Solvable non-deterministically in polynomial many steps on the size of the problem instances), if it is first efficient to guess a solution and then efficient to check that this solution is correct. It is conjectured (and not known) that $P \neq NP$. This is one of the great problems in modern mathematics and theoretical computer science. If the guessing in *NP*-problems can be replaced by an efficient systematic search for a solution, then $P = NP$. For any one *NP*-complete problem, if it is in *P*, then necessarily $P = NP$.

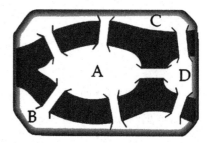

Fig. 4.1 The town of Konigsberg and its seven bridges

4.3   Graphs and Their Plane Figures

Let V be a finite set, and denote by $E(V) = \{\{u, v\} \mid u, v \in V, u \neq v\}$ the subsets of V of two distinct elements. A pair $G = \{V, E\}$ with $E \subseteq E(V)$ is called a graph (on V). The elements of V are the vertices, and those of E the edges of the graph. The vertex set of a graph G is denoted by V_G and its edge set by E_G. Therefore $G = \{V_G, E_G\}$. In literature, graphs are also called simple graphs; vertices are called nodes or points; edges are called lines or links. A pair $\{u, v\}$ is usually written simply as uv. Notice that then $uv = vu$. In order to simplify notations, we also write $v \in G$ instead of $v \in V_G$. For a graph G, we denote $v_G = |V_G|$ and $\varepsilon_G = |E_G|$. The number v_G of the vertices is called the order of G, and ε_G is the size of G. For an edge $e = uv \in E_G$, the vertices u and v are its ends. Vertices u and v are adjacent or neighbors, if $e = uv \in E_G$. Two edges $e_1 = uv$ and $e_2 = uw$ having a common end, are adjacent with each other.

A graph G can be represented as a plane figure by drawing a line between the points u and v (representing vertices) if $e = uv$ is an edge of G. Figure 4.2 is a drawing of the graph G with $V_G = \{v_1, v_2, v_3, v_4, v_5, v_6\}$ and $E_G = \{v_1v_2, v_1v_3, v_2v_3, v_2v_4, v_5v_6\}$.

4.3.1   Direct Graph

A graph where the edges have a direction, that is, the edges are ordered, directed graphs or digraphs $D = \{(V, E), E \subseteq V \times V$. In this case, $uv \neq vu\}$.

The directed graphs have representations, where the edges are drawn as arrows as shown in Fig. 4.3. A digraph can contain edges uv and vu of opposite directions.

4.3.2   Undirected Graph

A graph for which the relations between pairs of vertices are symmetric, so that each edge has no directional $UD = \{(V, E), E \subseteq V \times V$. In this case, $uv = vu\}$. Figure 4.4 shows an undirected graph.

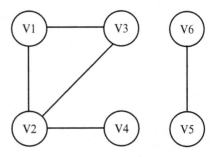

Fig. 4.2 The graph G with $V_G = \{v_1, v_2, v_3, v_4, v_5, v_6\}$ and $E_G = \{v_1v_2, v_1v_3, v_2v_3, v_2v_4, v_5v_6\}$

Fig. 4.3 The direct graph

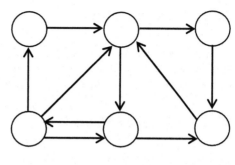

Fig. 4.4 The undirected graph

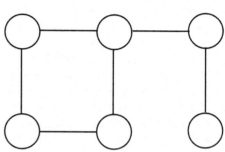

4.3.3   Network Representations

Networks are mathematically represented by graphs where vertexes denote nodes of the network and edges of the graph are the existing communication links between nodes. Thus a proper definition of graphs and their mathematical representations, together with a few theoretical results, will be useful in the study and classification of networks and particularly in the study of distributed algorithms.

A graph $G = \{V, E\}$ is defined as a pair, where V is a finite set of vertexes and E a set of edges. The set of edges E is a subset of the set $V \times V$ of ordered pairs of distinct vertexes.

4.3.4   Node Degree

For an undirected graph the *degree of a node* is equal to the number of incident edges on that node. Particularly in the case of loops they count as two, given that a loop has a leaving and entering end of the same edge on the node.

4.3.5   Distance Matrix

A symmetric N by N matrix in which elements M_{ij} represent the length of shortest path between i and j; if there is no such path $M_{ij} = \infty$. It can be derived from powers of the Adjacency matrix.

4.3.6   Incidence Matrix

The incidence matrix of a directed graph G is a $p \times q$ matrix $[b_{ij}]$ where p and q are the number of vertexes and edges respectively, such that $b_{ij} = 1$ if the edge x_j leaves vertex v_i, -1 if it enters vertex v_i and 0 otherwise.

4.3.7   Adjacency Matrix

The adjacency matrix of a finite directed or undirected graph G on n vertexes is the $n \times n$ matrix where the nondiagonal entry a_{ij} is the number of edges from vertex i to vertex j, and the diagonal entry a_{ii} is either twice the number of loops at vertex i or just the number of loops (usages differ, depending on the mathematical needs; the former convention is normally used for undirected graphs, though directed graphs always follow the latter). There exists a unique adjacency matrix for each graph (up to permuting rows and columns), and it is not the adjacency matrix of any other graph. In the special case of a finite simple graph, the adjacency matrix is a (0,1)-matrix with zeros on its diagonal.

If the graph is undirected, the adjacency matrix is symmetric. In this project we will assume undirected graphs in the representation of networks from the assumption that communication is bidirectional between nodes.

We define the $N \times N$ adjacency matrix, A, as

$$A_{i,j} = \begin{cases} 1 & \textit{if } (i,j) \in E \\ 0 & \textit{otherwise} \end{cases} \tag{4.1}$$

4.3.8   Degree Matrix

It is a diagonal matrix which contains information about the degree of each vertex. I.e., given a graph $G = \{V, E\}$ with $\|V\| = n$ the degree matrix D for G is a $n \times n$ square matrix defined as

$$d_{i,j} = \begin{cases} \deg(v_i) & \textit{if } i = j \\ 0 & \textit{otherwise} \end{cases} \tag{4.2}$$

or more compactly $D = \text{diag}(A.\mathbf{1})$.

4.3.9   Laplacian Matrix

The Laplacian of a graph G is defined as:

$$L = D - A \tag{4.3}$$

With D the degree matrix of G and A the adjacency matrix of G.
More explicitly, given a graph G with n vertexes, the matrix L satisfies

$$l_{i,j} = \begin{cases} \deg(v_i) & \text{if } i = j \\ -1 & \text{if } i \neq \text{ and } v_i \text{ adjacent } v_j \\ 0 & \text{otherwise} \end{cases} \tag{4.4}$$

In the case of directed graphs, either the in-degree or the out-degree might be used, depending on the application. Spectral properties of Laplacian matrix will play an essential role in analyzing the convergence of the class of linear consensus protocols. According to Gershgorin theorem, all eigenvalues of L in the complex plane are located in a closed disk centered at $\Delta + 0_j$ with a radius of $\Delta = max_i\ d_i$, which is the *maximum degree* of a graph. For undirected graphs, L is a symmetric matrix with real eigenvalues and the set of eigenvalues of L can be ordered in an ascending order as

$$0 = \lambda_1 \leq \lambda_2 \leq \ldots \ldots \lambda_n \leq 2\Delta \tag{4.5}$$

The zero eigenvalue is the trivial eigenvalue of L and its multiplicity is the number of connected components of G. For a connected graph G, $\lambda_2 > 0$. The second smallest eigenvalue of Laplacian, λ_2, is called *algebraic connectivity* of a graph. Algebraic connectivity of the network topology is a criterion for the speed of convergence of consensus algorithms.

The interaction topology of a network is represented using a directed graph $G = \{V, E\}$ is a pair, where V is a finite set of vertexes $V = \{1, 2, \ldots, n\}$ and E a set of edges E is a subset of the set $V \times V$ of ordered pairs of distinct vertexes. The neighbors of agent i are denoted by $N_i = \{j \in V : (i, j) \in E\}$.

4.4 Central Kalman Filter in Wireless Sensor Network

Let's consider a sensor network with n sensors that are interconnected via an undirected graph. Let's consider a sensor network with n sensors that are interconnected via an undirected graph. The model of a process can be defined as:

$$x_{k+1} = A_k x_k + B_k u_k + w_k \qquad k \geq 0 \tag{4.6}$$

$$z_k = C_k x_k + v_k \qquad k \geq 0 \tag{4.7}$$

where $z_k \in R^{np}$ represents the vector of p-dimensional measurements obtained via n sensors, w_k and v_k are assumed to be $m \times 1$ and $p \times 1$ zero-mean white noise processes, respectively. The process v_k is called measurement noise and w_k is called process noise.

The above equations have several variables:

1. A, B, C: System matrices
2. k: The time index
3. x: The state system
4. u: Input to the system
5. z: The measurement output
6. w and v: Process noise and measurement noise respectively and they are zero mean mutually uncorrelated white noises with covariance's,

$$E\left[W_k \ W_k^T\right] = Q_k \tag{4.8}$$

$$E\left[V_k \ V_k^T\right] = R_k \tag{4.9}$$

Additionally, $x_0 \in R^m$ is the zero-mean initial state of the process with covariance matrix P_0, and is assumed to be uncorrelated with u_k and v_k.

Building upon the underlying dynamic system model, and introducing measurements $Z_k = \{z_0, z_1, \ldots, z_k\}$, we define the information matrix to be the inverse of the state covariance matrix. To gain some intuition behind this definition, lets for sake of practicality think of the state covariance matrix and the information matrix as scalar quantities. Then, invoking the limits of zero and infinity for the covariance, we can think of the information as achieving its maximum when the covariance is zero, and vice versa.

Thus, covariance is essentially a measure of how close our estimate is to the true value, the construction of the information matrix makes sense. The higher the covariance, the less amount of information is contained in the estimate.

The Kalman filter estimates a process by using a form of feedback control. The filter estimates the process state at some time and then obtains feedback in the form of measurements. The equations for the Kalman filter fall into two groups:

1. Time update equations
2. Measurement update equations

The time update equations are responsible for projecting forward the current state and error covariance estimates to obtain the a priori estimates for the next time step. The measurement update equations are responsible for the feedback, i.e. for incorporating a new measurement into a priori estimate to obtain an improved a posteriori estimate.

The time update equations can also be thought of as predictor equations, while the measurement update equations can be thought of as corrector equations. Indeed the final estimation algorithm resembles that of a predictor-corrector algorithm for solving numerical problems as shown below in Fig. 4.5.

Following from the above discussion, the inverses of the state covariance matrices P and M are known as the information matrices. We describe below the Kalman filter iterations in the information form.

Fig. 4.5 The Kalman filter
cycle

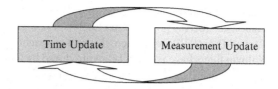

1. Measurement-update:

$$M_k^{-1} = P_k^{-1} + C_k^* R_k^{-1} C_k \qquad (4.10)$$

2. Time-update: We will employ a change of notation here, for reasons that will follow naturally given the dynamics of the following equations. For state estimates and updates we define:

$$K_k = M_k C_k^* R_k^{-1} \qquad (4.11)$$

$$\hat{x}_{(k|k)} = \hat{x}_{(k|k-1)} + K_k(z_k - C_k \hat{x}_{(k|k-1)}) \qquad (4.12)$$

$$P_{k+1} = A_k M_k A_k^* + B_k Q_k B_k^* \qquad (4.13)$$

$$\hat{x}_{(k+1|k)} = A_k \hat{x}_{(k|k)} \qquad (4.14)$$

where Q_k is the covariance of the process noise w_k and R_k is covariance of the measurement noise v_k. The calculation of Kalman gain (K_k) not only depends on the known measurement error covariance R_k, but also the state estimation covariance P_k.

Thus far we have described in detail the workings of a central Kalman filter in the context of a sensor network with n nodes, where each node observes p various measurements. In central Kalman filter, all the observations of the sensors are passed back to a central processing facility to perform overall data fusion so they suffer from the associated problems: potential computational bottlenecks and the susceptibility to total system failure if the central facility should fail. Thus central Kalman filter is not applicable to be implemented in wireless sensor network.

4.5 Distributed Kalman Filter (DKF) Literature Work

Consensus filters can be used independently for DKF. The role of this consensus filter is to perform distributed fusion of sensor measurements that is necessary for implementation of a scalable Kalman filter. Consensus problems and their special cases have been the subject of intensive studies by several researchers [6–14] in the context of formation control, self-alignment, and flocking [15] in networked dynamic systems. Consensus-based tracking [16, 17] and synchronization algorithms [18] in sensor networks that are scalable and resilient have recently emerged as powerful tools for collaborative information processing. Distributed Kalman filter algorithms rely on consensus filters is proposed in [19, 20]. A new generation of DKF algorithms with a Peer-to-Peer (P2P) architecture that rely on reaching a consensus on estimates of local Kalman filters have recently been introduced by Olfati-Saber in [19] they refer to this class of distributed estimation algorithms as Kalman-Consensus Filters (KCF). Carli studied the interaction between the consensus matrix and the number of messages exchanged per sampling time in the Kalman filter for scalar systems. They proved that optimizing the consensus matrix for fastest convergence and using the centralized optimal gain is not necessarily the optimal strategy if the number of exchanged messages per sampling time is small [22].

The consensus problem with quantized transmission has been studied recently. Xiao in [23, 24] studied the convergence of the model in [6] when the received values are assumed corrupted with an additive noise, and show that the variance of the state vector with respect to the average of the initial values diverges with time. Schizas in [25] proposes a distributed MLE and BLUE estimators for the estimation of deterministic signals in ad hoc WSNs, where the estimators are formulated as the solution of convex minimization sub problems. Kashyap in [26] introduced the concept of quantized consensus and propose an algorithm to reach a consensus in that sense. Some contributions found in literature analyze the communication bandwidth constrains. Huimin in [27] studied the tradeoff between bandwidth and tracking accuracy with communication constraints. Ribeiro in [28] analyzed the distributed state estimators of dynamical stochastic processes, whereby the low communication cost is affected by requiring the transmission of a single bit per observation. In [29] Olfati-Saber addressed the distributed tracking of a maneuvering target using sensor networks with limited sensing range. They introduce a novel switching model for a target that is able to remain inside a rectangle in all time. Carli in [22] introduced the problem of estimating the state of a dynamical system from distributed noisy measurements. Each agent builds a local estimate based on its own measurements and estimates from its neighbors. Estimation is performed via a two stage strategy, the first is a Kalman-like measurement update which does not require communication, and the second is an estimate fusion using a consensus matrix. They studied the interaction between the consensus matrix, the Kalman gain, and the number of messages exchange per sampling time. They proved that optimizing the consensus matrix for fastest convergence and using the centralized optimal gain is not necessarily the optimal strategy if the number of message exchange per sampling time is small. Moreover, they verified that under certain conditions the optimal consensus matrix should be doubly stochastic.

4.6   Olfati-Saber's Distributed Kalman Filter

Olfati-Saber introduced a novel distributed Kalman filter strategy for distributed state estimation and target tracking in sensor networks [6, 7, 9, 15–17, 21, 29, 30]. This DKF strategy consists of identical high-pass consensus filters for distributed fusion of sensor data and covariance information. It enables a sensor network to act as a collective observer for a process that is not observable by an individual sensor. The strategy only requires single-hop communication between a sensor and its neighbors.

4.7 Consensus Filters

Consensus problems are widely considered in computer science and they have a long history in this field. They basically formed the field of distributed computing. Formal study of these types of problems goes back to people who were working in management science and statistics in the 1960s. The notion of statistical consensus theory by DeGroot attracted interest 20 years later in the problem of processing information with uncertainty obtained from multiple sensors and medical experts.

Distributed computing has been considered by people in systems and control theory starting with the work of Olfati-Saber, and John [31] on asynchronous asymptotic agreement problem for distributed decision-making systems.

In a network, consensus means to get an agreement regarding some common interest of the nods which depends on the states of all them. A consensus algorithm is the law which specifies the information flow between and node and its neighbors to reach to the consensus in the whole network.

A directed graph $G = \{V, E\}$ with the set of nodes $V = 1, 2, \ldots, n$ and edges $E \subseteq V \times V$ is employed to show the interaction between the nodes in a network. The neighbors of the node i are denoted by the set $\{N_i = j \in V: (i, j) \in E\}$. A simple consensus protocol, to reach a consensus on a graph regarding the state of n integrator nodes with dynamics $\{\dot{x}_i = u_i\}$, can be expressed as an nth-order linear system:

$$\dot{x}_i(t) = \sum_{j \in N_i} \left(x_j(t) - x_i(t) + b_i(t) \right), \ x_i(0) = z_i \ \in R \ b_i(t) = 0 \qquad (4.15)$$

We can collect the terms in 4.6 and rewrite it as

$$\dot{x} = -Lx \qquad (4.16)$$

where $L = [l_{ij}]$ is the graph Laplacian of the network.

According to the definition of graph Laplacian, all row-sums of L are zero since $\sum_j l_{ij} = 0$. Therefore, L always has a zero eigenvalue $\lambda_1 = 0$. This zero eigenvalues correspond to the eigenvector $\mathbf{1} = (1, \ldots, 1)^T$ because $\mathbf{1}$ belongs to the null-space of L, in other words $L\mathbf{1} = 0$. So, we can conclude that an equilibrium of system is a state in the form $x^* = (\alpha, \ldots, \alpha)^T = \alpha\mathbf{1}$, where all nodes get to a consensus. Using some analytical tools from algebraic graph theory, we later show that x^* is a unique equilibrium for connected graphs.

4.7.1   Information Consensus in Networked Systems

Let us consider a network of agents whose dynamics is $\dot{x}_i = u_i$. Their goal is to achieve a consensus through communication with their neighbors on a graph $G = (V, E)$. By reaching a consensus we mean all the agents get to the same state value, i.e. their states satisfy the following equation:

$$x_1 = x_2 = \cdots = x_n \tag{4.17}$$

This consensus value usually is called agreement space and can be expressed as $x = \alpha\mathbf{1}$ where $\alpha \in R$ is the collective decision of the group of nodes. Let $A = [a_{ij}]$ be the adjacency matrix of graph G. The set of neighbors of node i is N_i and defined as:

$$N_i = \{j \in V : \ a_{ij} \neq 0\}; \ V = \{1, 2, \ldots \ldots \ldots, n\} \tag{4.18}$$

Node i communicates with node j if j belongs to the neighbor set of i. The set of all nodes and their neighbors defines the edge set of the graph $E = \{(i, j) \in VXV : \ a_{ij} \neq 0\}$.

A distributed consensus algorithm guarantees convergence to a collective decision via local interconnection between nodes. Assuming the graph is undirected, i.e. $a_{ij} = a_{ji}$ for all i, j, it follows that the state of all nodes is an invariant quantity, or $\sum_j \dot{x} = 0$. Applying this condition at times $t = 0$ and $t = \infty$ leads to the following result:

$$\alpha = \frac{1}{n}\sum_i x_i(0) \tag{4.19}$$

Thus, if a consensus is achieved, the collective decision is equal to the average of the initial state of all the nodes. A consensus protocol with the mentioned invariance property is called an average-consensus algorithm.

1. Low-Pass consensus filter: Let us assume that there is a network with n nodes, x_i denote the m-dimensional state of node i and u_i denote the dimensional input of node i. Then, the following consensus protocol is a low-pass consensus filter:

$$\dot{x}_i = \sum_{j \in N_i} a_{ij}(x_j - x_i) + \sum_{j \in N_i \cup \{i\}} a_{ij}(u_j - x_i) \tag{4.20}$$

It can be expressed in the following collective form

$$\dot{x} = -\left(I_{mn} + \hat{D} + \hat{L}\right)x + \left(I_{mn} + \hat{D} - \hat{L}\right)u \tag{4.21}$$

where $x = [x_1, x_2, \ldots, x_n]^T$, $\hat{A} = A \oplus I_m$ and $\hat{L} = L \oplus I_m$. The MIMO transfer function of (4.21) from input u to output x is:

$$H_{lp}(s) = \left[sI_{mn} + \left(I_{mn} + \hat{D} + \hat{L}\right)^{-1} \left(I_{mn} + \hat{D} - \hat{L}\right) \right] \tag{4.22}$$

Applying Gregorian theorem to matrix $-\left(I_{mn} + \hat{D} + \hat{L}\right) = -\left(I_{mn} + 2D - \hat{A}\right)$ guarantees that all poles of $H(s)$ are strictly negative, and thus the filter is stable. Moreover their real part falls in the interval $[1(1 + 3d_{max}), -(1 + 3d_{min})]$, where $d_{max} = \max_i d_i$ and $d_{min} = \min_i d_i$. On the other hand $H(s)$ is a proper MIMO transfer function satisfying $lim_{s \to \infty} H(s) = 0$, which means that it is a low-pass filter.

2. High-pass consensus filter: Let us assume that there is a network with n nodes, x_i be the m-dimensional state of the node i and u_i be the dimensional input of this node. Then, the following dynamic consensus algorithm is a high-pass filter.

$$\dot{x}_i = \sum_{j \in N_i} \left(x_j - x_i\right) + \dot{u}_i \tag{4.23}$$

This relation can be re-stated as follows

$$\dot{x}_i = -\hat{L}x + \dot{u} \tag{4.24}$$

where $\hat{L} = L \oplus I_m$. The improper Multiple-Input and Multiple-Output (MIMO) transfer function of this high-pass consensus filter from input u to output x is:

$$H_{hp}(s) = \left(sI_{nm} + \hat{L}\right)^{-1} s \tag{4.25}$$

As we can see $lim_{s \to \infty} H(s) = I_{nm}$, which means that the filter propagates high frequency noise and is not useful for sensor fusion by itself.

3. Band-Pass consensus filter: This band-Pass distributed filter can be defined as:

$$H_{bp}(s) = H_{lp}(s) \, H_{hp}(s) \tag{4.26}$$

It has the following dynamics

$$\begin{aligned} \dot{x}_1 &= -\left(I_{mn} + \hat{A} + 2\hat{L}\right)x_1 + \left(I_{mn} + \hat{A}\right)u \\ \dot{x}_2 &= -\hat{L}x_2 + \dot{x}_1 \end{aligned} \tag{4.27}$$

With input u and output x_2.

4.7.2 Distributed Kalman Filter with Embedded Consensus Filters

Hereby, z_k was an np-dimensional vector, essentially a long vertical vector with stacked observations from the n different sensor. The process is an m-dimensional process i.e., $x_k \in R^m$ and the corresponding white noise vectors have appropriate dimensions matching the z_k and x_k. There are n various sensors; each sensor is m-dimensional meaning there are m different states associated with each sensor. And for each of those states, there are p measurements taken. The states are related to each other by way of matrices A and B. Likewise; our z_k is extracted from our x_k by means of a linear combination dictated by the matrix C.

We will start by rewriting, $z_k = C_k x_k + v_k$, the sensing model equation, which again is essentially equating two $np \times 1$ vectors. These vectors were stacked with the information obtained at each individual sensor. In the distributed scenario we will consider each individual sensor one at a time, producing the following equation:

$$z_k(k) = C_k(k) + v_k(k) \tag{4.28}$$

This differs from the original sensing model because it describes the activity for an individual sensor. This intuition is supported mathematically by the dimensions of variables. $z_i(k)$ now has dimensions $p \times 1$ instead of $np \times 1$. C_k now has dimension $p \times m$ instead of $np \times m$ and lastly $v_i(k)$ now has dimensions $p \times 1$ instead of $np \times 1$.

Now that we have developed a consistent notation for describing the state activity at each sensor, we can define new variables z_c, v_c, and C_c, that are nothing but a collection of the each parameter gathered from all nodes. Hence, there are n entries for each of these newly defined variables. We describe $z_c = col(z_1, z_2, \ldots, z_n)$, $v_c = col(v_1, v_2, \ldots, v_n)$, and $C_c = col(C_1, C_2, \ldots, C_n)$. This naturally results in the state relation: $z_c(k) = C_c(k) + v_c(k)$. Invoking the statistics from the white Gaussian noise perturbations, and again defining a variable R_c as a collection of covariance from the n various sensors, we can simply write $R_c = diag(R_1, R_2, \ldots, R_n)$. This definition allows us to express the Kalman filter iterations from the point of view of the central node. Notice that the following equations strongly resemble the Kalman filter iterations before a distributed implementation was considered. In this case we introduced the iterations at individual nodes, then combined them to arrive at the original set of Kalman filter iterations. The only difference is that the index now is c for central instead of the k index that was originally used to describe the set of measurements. We have the following:

$$M = \left(P^{-1} + C_c^* R_c^{-1} C_c\right)^{-1} \tag{4.29}$$

$$K_c = M \, C_c^* \, R_c^{-1} \tag{4.30}$$

$$\hat{x}_k = \hat{x}_{(k|k-1)} + K_c\left(z_c - C_c \hat{x}_{(k|k-1)}\right) \tag{4.31}$$

The distributed implementation of the Kalman filter relies on two consensus problems executed at each iteration. The first of these two is the determination of an $m \times m$ matrix defined as such:

$$S = \frac{1}{n} C_c^* \, R_c^{-1} C_c = \frac{1}{n} \sum_{i=1}^{n} C_i^* \, R_i^{-1} C_i \tag{4.32}$$

The second consensus determination is an m-vector of average measurements, where each measurement is defined as:

$$y_i = C_i^* R_i^{-1} z_i = \frac{1}{n} \sum_{i=1}^{n} y_i \tag{4.33}$$

With these two definitions in mind, and with the application of some clever arithmetic manipulations, the state propagation expressed above from the perspective of the central node can be rewritten as:

$$\hat{x}_k = \hat{x}_{(k|k-1)} + nM(y_i - S\hat{x}_{(k|k-1)}) \tag{4.34}$$

This is effectively the Kalman state update equation for each node. Upon close examination of this equation, it is natural to observe that the gain is the product nM. Remembering that M is the state covariance matrix and is given by $M = \left(P^{-1} + C_c^* R_c^{-1} C_c\right)^{-1}$, we can express nM in the following revealing manner:

$$nM = M_\mu = \left((nP)^{-1} + S\right)^{-1} \tag{4.35}$$

Let us summarize the above arguments for constructing the state update equations of a distributed Kalman filter. The aforementioned expressions for state equations and covariance matrices are placed in the context of a sensor network with n sensors and a topology G that is a connected graph illustrating a process of dimension m using $p \leq m$ sensor measurements. At each iteration k, every sensor solves two consensus problems, acquiring the parameter S and the parameter y. This enables each node to calculate the state estimate using the μ-Kalman filter update equations:

$$M_\mu = \left((nP)^{-1} + S\right)^{-1} \tag{4.36}$$

$$\hat{x}_{(k|k)} = \hat{x}_{(k|k-1)} + M_\mu(y - S\hat{x}_{(k|k-1)}) \tag{4.37}$$

$$P_\mu^+ = A M_\mu A^* + nBQB^* \tag{4.38}$$

$$\hat{x}_{(k+1|k)} = A_k \hat{x}_{(k|k)} \tag{4.39}$$

Figure 4.6 presents the architecture of a node running a μ-Kalman filter with embedded consensus, and the communication architecture between two nodes.

Of course, the most significant attribute of the above μ-Kalman filter state update equations lies in the fact that the state estimates produced are identical to the ones

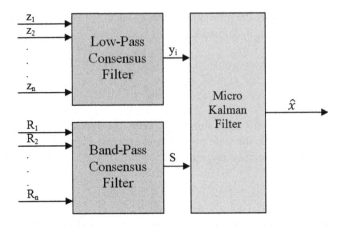

Fig. 4.6 Schematic representation of the μ-Kalman filter

obtained via a central Kalman filter. Furthermore, a significant advantage for the distributed implementation is revealed upon examining the computational costs of the respective gain matrices. The central Kalman filter gain K has $O\ (m^2 n)$ elements while the gain M_μ of the μ-Kalman filter has $O\ (m^2)$ elements. This suggests that the implementation of the μ-Kalman filter is in fact more computationally feasible than that of the central Kalman filter. A last consideration in the topic of μ-Kalman filters returns to their usage of the time-varying consensus values S and y. Because of the difference in nature between the two consensus values, two separate approaches will be taken to obtain the desired quantities. Specifically, the calculation of S is shown to require a type of band-pass filter, while y is obtained from a collection of node measurements, hence justifying the use of a low-pass filter. The time-varying nature of both parameters along with the mechanisms in place for estimation of these necessary parameters naturally leads to some error at each iteration.

In order to implement this DKF in wireless sensor network the communication and computation issues will come up, as in Olfati's technique two values, measurement and covariance, need to be exchanged between node and this will consume the communication bandwidth. Moreover Kalman filter is a power hungry algorithm in term of computational complexity.

In this thesis we will offer solutions for these problem, Starting with the computational complexity problem, we proposed a novel light-weight and low-power multiplication algorithm. The proposed algorithm aims to decrease the number of instruction cycles, save power and reduce the memory storage without increasing the code complexity or sacrificing accuracy. For the communication bandwidth problem, we will exchange the estimates between nodes which will lead to save the communication bandwidth. Moreover, we proposed a DKF based on a well known fast polynomial filter to accelerate distributed average consensus in static network topologies. The idea is to apply a fast polynomial filter on the network matrix that will shape its spectrum in order to increase the convergence rate by minimizing its second largest eigenvalue. Fast convergence can contribute to significant energy saving and hence a fast DKF.

Bibliography

1. R.E. Kalman, "A new approach to linear filtering and prediction problems," *Journal of Basic Engineering,* vol. 3, pp. 35–45, 1960.
2. J. Speyer, "Computation and transmission requirements for a decentralized linear-quadratic-Gaussian control problem," *IEEE Transactions on Automatic Control,* vol. 24, pp. 266–269, May 1979.
3. B.S.Y. Rao, H.F. Durrant-Whyte, and J.A. Sheen, "A fully decentralized multi-sensor system for tracking and surveillance," *Journal of Robotics Research,* vol. 3, pp. 20–44, November 1993.
4. J. Dong, Q. Chen, and Z. Niu, "Random graph theory based connectivity analysis in wireless sensor networks with Rayleigh fading channels," in *Proceeding of the Asia-Pacific Conference on Communications,* Hong Kong, China, March 2007, pp. 123–126.
5. B. Bollabas, *Random Graphs Second Edition*: Cambride University Press, 2001.
6. R. Olfati-Saber and R.M. Murray, "Consensus problems in networks of agents with switching topology and time-delays," *IEEE Transactions on Automatic Control,* vol. 49, pp. 1520–1533, January 2004.
7. R. Olfati-Saber and R.M. Murray, "Consensus protocols for networks of dynamic agents," in *Proceeding of the American Control Conference,* Denver, Colorado USA, June 2003, pp. 951–956.
8. M. Mesbahi, "On State-dependent dynamic graphs and their controllability properties," *IEEE Transactions on Automatic Control,* vol. 50, pp. 387–392, May 2005.
9. R. Olfati-Saber, "Ultrafast consensus in small-world networks," in *Proceeding of the American Control Conference,* pp. 2371–2378, 2005.
10. Y. Hatano and M. Mesbahi, "Agreement over random networks," *IEEE Transactions on Automatic Control,* vol. 50, pp. 1867–1872, November 2005.
11. L. Moreau, "Stability of multiagent systems with time-dependent communication links," *IEEE Transactions on Automatic Control,* vol. 50, pp. 169–182, February 2005.
12. R. Wei and R. W. Beard, "Consensus seeking in multiagent systems under dynamically changing interaction topologies," *IEEE Transactions on Automatic Control,* vol. 50, pp. 655–661, November 2005.
13. L. Xiao and S. Boyd, "Fast linear iterations for distributed averaging," *Systems and Control Letters,* pp. 65–78, June 2004.
14. L. Xiao, S. Boyd, and S. Lall, "A scheme for robust distributed sensor fusion based on average consensus," in *the 4th International Symposium on Information Processing in Sensor Networks,* pp. 63–70, 2005.
15. R. Olfati-Saber, "Flocking for multi-agent dynamic systems: algorithms and theory," *IEEE Transactions on Automatic Control,* vol. 51, pp. 401–420, April 2006.
16. R. Olfati-Saber, "Distributed Kalman Filter with Embedded Consensus Filters," in *the 44th IEEE Conference on Decision and Control,* pp. 8179–8184, 2005.
17. R. Olfati-Saber, "Distributed tracking for mobile sensor networks with information-driven mobility," in *Proceeding of the American Control Conference,* New York City, USA , July 2007, pp. 4606–4612.
18. G. Scutari, S. Barbarossa, and L. Pescosolido, "Distributed Decision Through Self-Synchronizing Sensor Networks in the Presence of Propagation Delays and Asymmetric Channels," *IEEE Transactions on Signal Processing,* vol. 56, pp. 1667–1684, May 2008.
19. R. Olfati-Saber and J. S. Shamma, "Consensus Filters for Sensor Networks and Distributed Sensor Fusion," in *Proceeding of the 44th IEEE Conference on Decision and Control,* Seville, Spain, December 2005, pp. 6698–6703.
20. D. Spanos, R. Olfati-Saber, and R.M. Murray, "Dynamic Consensus on Mobile Networks," in *Proceeding of the 16th IFAC World Congress,* Prague, Czech Republic, July 2005, pp. 139–145.

21. R. Olfati-Saber, "Distributed Kalman filtering for sensor networks," in *Decision and Control, 2007 46th IEEE Conference on*, New Orleans, Louisiana, USA, December 2007, pp. 5492–5498.
22. R. Carli, A. Chiuso, L. Schenato, and S. Zampieri, "Distributed Kalman filtering based on consensus strategies," *IEEE Journal on Selected Areas in Communications*, vol. 26, pp. 622–633, November 2008.
23. L. Xiao, S. Boyd, and S.J. Kim, "Distributed average consensus with least-mean-square deviation," in *Proceeding of the 17th International Symposium on Mathematical Theory of Networks and Systems*, Kyoto, Japan, July 2006, pp. 2768–2776.
24. L. Xiao, S. Boyd, and S.-J. Kim, "Distributed average consensus with least–mean–square deviation," *Journal of Parallel and Distributed Computing*, vol. 2, pp. 33–46, May 2007.
25. I. D. Schizas, A. Ribeiro, and G.B. Giannakis, "Consensus in Ad Hoc WSNs With Noisy Links Part I: Distributed Estimation of Deterministic Signals," *IEEE Transactions on Signal Processing* vol. 56, pp. 350–364, June 2008.
26. A. Kashyap, T. Basar, and R. Srikant, "Quantized Consensus," in *Proceeding of the IEEE International Symposium on Information Theory*, pp. 635–639, 2006.
27. C. Huimin and X.R. Li, "On track fusion with communication constraints," in *Proceeding of the 10th International Conference on Information Fusion*, Seattle, Washington, USA, July 2007, pp. 1–7.
28. A. Ribeiro, G.B. Giannakis, and S.I. Roumeliotis, "SOI-KF: Distributed Kalman Filtering With Low-Cost Communications Using the Sign of Innovations," *IEEE Transactions on Signal Processing*, vol. 54, pp. 4782–4795, October 2006.
29. R. Olfati-Saber and N.F. Sandell, "Distributed tracking in sensor networks with limited sensing range," in *Proceeding of the American Control Conference*, Seattle, Washington, USA, June 2008, pp. 3157–3162.
30. R. Olfati-Saber, "Kalman-Consensus Filter : Optimality, stability, and performance," in *Proceeding of the 48th IEEE Conference on Decision and Control*, Chinghai, China, December 2009, pp. 7036–7042.
31. N.T. John and A. Michael, "Convergence and asymptotic agreement in distributed decision problems," in *Proceeding of the 21st IEEE Conference on Decision and Control*, Orlando, Florida, USA, December 1982, pp. 692–701.

Chapter 5
Proposed Distributed Kalman Filter

Abstract In this chapter, the DKF problem is addressed by reducing it into a dynamic consensus problem in term of weighted average estimates matrix that can be viewed as data fusion problem. We have presented a Distributed Kalman Filter based on polynomial filter to accelerate the distributed average consensus in the static network topologies. The proposed algorithm performs closely to the central filter, and also reduces the filter complexity at each node by reducing the dimension of the data. Thus, it scales computational complexity. Being based on sending only the estimates between neighbors, it also reduced radically the communication requirements. The proposed DKF contributes to significant energy saving.

5.1 Distributed Kalman Filter (DKF) in WSN and Related Work

Wireless sensor networks have become a widely used technology for applications ranging from military surveillance to industrial fault detection. So far, the development in micro-electronics has made it possible to build networks of inexpensive nodes characterized by modest computation and storage capability as well as limited battery life. A fundamental problem in wireless sensor networks is to solve detection and estimation problems using scalable algorithms i.e., dada fusion algorithm. This requires development of novel distributed algorithms for estimation and in particularly Kalman filter.

Consensus algorithm provides a scalable algorithm for wireless sensor fusion. This consensus filter plays a crucial role in solving a data fusion problem that allows implementation of a scheme for distributed Kalman filter in sensor networks. Consensus algorithms have proven to be effective tools for performing network-wide distributed computation tasks such as computing aggregate quantities and functions over networks. Consensus filters allow the network to agree on the value of a particular computation. In that sense one could think of

A. Abdelgawad and M. Bayoumi, *Resource-Aware Data Fusion Algorithms*
for Wireless Sensor Networks, Lecture Notes in Electrical Engineering 118,
DOI 10.1007/978-1-4614-1350-9_5, © Springer Science+Business Media, LLC 2012

two possible options: one, do an estimation on each of the nodes and then agree on an average value of all the nodes estimates; or one could get to a consensus on certain computation values, dependent on all the measurements of the network needed to calculate the estimate. The role of this consensus filter is to perform distributed fusion of sensor measurements that is necessary for implementation of a scalable Kalman filter.

Distributed average consensus is the task of calculating the average of a set of measurements made at different locations through the exchange of local messages. The goal is to evade the need for complicated networks with routing protocols and topologies and ensure that the final average is available at every node. Distributed average consensus algorithm is attractive because it obviates the need for global communication and complicated routing. Moreover it is robust to node and link failure. Distributed average consensus algorithm has been considered by people in systems and control theory starting with the work of John [1] on asynchronous asymptotic agreement problem for distributed decision-making systems. More recently, it has been applied in distributed coordination of mobile autonomous agents [2] and distributed data fusion in sensor networks [3–8].

In distributed average consensus algorithm, each node initializes its state to the local measurement at each iteration of the algorithm and then updates its state by adding a weighted sum of the local nodes. It is time-independent, and the state values converge to the average of the measurements asymptotically. Moreover, it is attractive because it is completely distributed and the computation at each node is very simple. The major deficiency is the relatively slow rate of convergence towards the average; often many iterations are required before the majority of nodes have a state value close to the average.

The convergence rate of distributed average consensus algorithms has been studied by several authors [9, 10]. Xiao, Boyd and their collaborators have been the main contributors of methods that strive to accelerate consensus algorithms through optimization of the weight matrix [3, 9, 11]. They showed that it is possible to formulate the problem of identifying the weight matrix that satisfies network topology constraints and minimizes the asymptotic convergence time as a convex semidefinite optimization task. This can be solved using a matrix optimization algorithm. Although elegant, the approach has two disadvantages. First, the convex optimization requires substantial computational resources and can impose delays in configuration of the network. If the network topology changes over time, this can be of particular concern. Second, a straightforward implementation of the algorithm requires a fusion centre that is aware of the global network topology. In particular, in the case of on-line operation and a dynamic network topology, the fusion centre needs to re-calculate the optimal weight matrix every time the network topology changes. If such a fusion centre can be established in this situation, then the value of a consensus algorithm becomes questionable. To combat the second problem, Boyd proposes the use of iterative optimization based on the sub-gradient algorithm. Calculation of the sub-gradient requires knowledge of the eigenvector corresponding to the second largest eigenvalue of the weight matrix. In order to make the algorithm distributed, Boyd employs decentralized orthogonal iterations [11] for eigenvector calculation. The resulting algorithm, although distributed, is demanding in terms of time, computation and communication, because it essentially involves two consensus procedures.

Sundaram achieved consensus in a finite number of time steps, and constituted an optimal acceleration for some topologies [12]. The disadvantage of the approach is that each node must know the complete weight matrix, retain a history of all state values, and then solve a system of linear equations. Again, this disadvantage is most consequential in the scenario where nodes discover the network online and the topology is dynamic, so that the initialization operation must be performed frequently. However, even in the simpler case of a static topology, the overhead of distributing the required initialization information can diminish the benefits of the consensus algorithm unless it is performed many times. Cao proposes an acceleration framework for gossip algorithms observing their similarity to the power method [13]. This framework is based on the use of the weighted sum of shift registers storing the values of local gossip iterations.

In this chapter, our goal is to accelerate the distributed Kalman filter in a fixed network. We apply a fast polynomial filter methodology in the weight matrix in order to accelerate the distributed average consensus in the network. The mean idea behind the polynomial filter is to shape the spectrum of the polynomial weight matrix to minimize the second largest eigenvalue and subsequently increase the convergence rate.

5.2 Network Representations

In wireless sensor network, there is a link between two nodes when packets can be successfully delivered from one node to the other. A wireless sensor network is called connected if for two arbitrary nodes, there is a route, which consists of such links, from one to the other. Traditional work on connectivity analysis of wireless sensor networks often focuses on finding a critical transmission range to keep the network connected. However, some low-cost sensor nodes may not support power-adaptive transmissions. On the other hand, changing the transmission range can be reformulated as changing the density of the sensor networks, in which each node is using fixed transmitted power [14]. Recently, random graph theory was introduced into the modeling of sensor networks with uncertain features. A random graph often can be imagined as a living organism which evolves with time. By giving a set of vertices in advance, the edges are generated according to some randomization rules [15].

Let us consider a static network topology, where the state of a link does not changes over the iterations. We assume the network at any arbitrary iteration t as an undirected graph $G = \{V, E\}$ with the set of nodes $V = 1, 2, \ldots, n$ and E is the edge set at iteration t. $E \subseteq E^*$, where $E^* \subseteq V \times V$ is employed to show the interaction between the nodes in a network and it is drawn if and only if sensor i can communicate with sensor j. The neighbors of the node i are denoted by the set $\{N_i = j \in V : (i, j) \in E\}$.

5.3 Asymptotic Average Consensus with Polynomial Filter

Let $x_0(i)$ be a real scalar assigned to node i at time $t = 0$ and $x_0 = (x_0(1), \ldots, x_0(n))$ denotes the vector of the initial values on the network. The distributed average consensus problem is to compute the average $\left(\frac{1}{n}\sum_{i=0}^{n} x_0(i)\right)$ at every node, via local communication and computation with their neighbors only on the graph. Thus, node i carries out its update, at each step, based on its local state and communication with its neighbors.

To reach a consensus on a graph, each sensor node i reports a scalar value $x_0(i) \in \mathbf{R}$. The vector of initial values on the network x_0 is denoted by

$$x_0 = [x_0(1),\ x_0(2),\ \ldots,\ x_0(n)]^T \in \mathbf{R}^n \tag{5.1}$$

The purpose of the consensus algorithm is to compute the average at each sensor node using linear distribution iteration. Thus the distributed linear iterations of the network can be defined in the following form:

$$x_{t+1}(i) = W_{ii}x_t(i) + \sum_{j \in N_i} W_{ij}x_t(j) \tag{5.2}$$

where $i = 1, \ldots, n$, $x_t(j)$ is the value computed by sensor node j at iteration t, and W_{ij} represents the edge weights of G. Each sensor node communicates only with its direct neighbors, so $W_{ij} = 0$. Writing in a matrix-vector format, the above update equation becomes

$$x_{t+1} = W^t x_t \tag{5.3}$$

where W is the weight matrix corresponding to the graph G of iteration t. The iterative relation given by Eq. 5.3 can be written as:

$$x_t = \left(\prod_{i=0}^{t-1} W^t\right) x_0 \tag{5.4}$$

Equation 5.4 means that $x_t = W^t x_0$ for all t. We want to chose the weight matrix W so that for any initial value x_0, x_t convergences to the average vector $\bar{x} = \left(\left(\frac{1}{n}\right)1^T x_0\right)1 = \left(\frac{1}{n}\right)11^T x_0$ i.e.,

$$\lim_{t\to\infty} x_t = \lim_{t\to\infty} W^t x_0 = \left(\frac{1}{n}\right)11^T x_0 \tag{5.5}$$

where 1 is the vector of ones. This is equivalent to the matrix equation:

$$\lim_{t \to \infty} W^t = \left(\frac{1}{n}\right) 11^T x_0 \tag{5.6}$$

Let us define the vector of average as:

$$\bar{x} = \frac{1}{n} \sum_{i=1}^{n} x_0(i) \tag{5.7}$$

From Eqs. 5.6 and 5.7, we can find:

$$\lim_{t \to \infty} x_t = \left(\left(\frac{1}{n}\right) 11^T\right) x_0 = \bar{x} \, 1 \tag{5.8}$$

The asymptotic convergence factor is defined as:

$$r_{asym}(W) = \underset{x_0 \neq x}{SUP} \lim_{t \to \infty} \left(\frac{\| x_t - \bar{x} \|_2}{\| x_0 - \bar{x} \|_2}\right)^{\frac{1}{t}} \tag{5.9}$$

Equation 5.6 hold if and only if

$$1^T W = 1^T \tag{5.10}$$

$$W1 = 1 \tag{5.11}$$

$$\rho\left(W - \left(\frac{1}{n}\right) 11^T\right) < 1 \tag{5.12}$$

where $\rho(.)$ is the spectral radius of a matrix. Now Eq. 5.9 can be written as:

$$r_{asymp}(W) = \rho\left(W - \left(\frac{1}{n}\right) 11^T\right) \tag{5.13}$$

Since W is symmetric, its eigenvalues arranges as: $\lambda_1(W) \geq \lambda_2(W) \ldots \geq \lambda_n(W)$. $\lambda_2(W)$, the second largest eigenvalue, is a measure of performance/speed of consensus algorithm [16, 17]. Thus, the convergence rate of Eq. 5.3 depends on the magnitude of the second largest eigenvalue λ_2.

5.4 Proposed Distributed Kalman Filter

We have described in detail the workings of a central Kalman filter in the context of a sensor network with n nodes, where each node observes p various measurements. The process we are describing is m-dimensional processes i.e. $x_k \in \mathbf{R}^m$ and the corresponding white noise vectors have appropriate dimensions matching the z_k and x_k. There are n various sensors; each sensor is m-dimensional, meaning there are

m different states associated with each sensor and for each of those states, there are p measurements taken. The states are related to each other by way of matrices A and B. Likewise; z_k is extracted from x_k by means of a linear combination dictated by the matrix C.

In the distributed scenario we will consider each individual sensor one at a time. The expressions for state equations and covariance matrices are placed in the context of a sensor network with n sensors and a topology G that is a connected graph illustrating a process of dimension m using $p \leq m$ sensor measurements. At each iteration k, each sensor node calculates the state estimate using the μ-Kalman filter update equations:

$$M_k^{-1} = P_k^{-1} + C_k^* R_k^{-1} C_k \tag{5.14}$$

$$K_k = M_k\, C_k^*\, R_k^{-1} \tag{5.15}$$

$$P_{k+1} = A_k M_k A_k^* + B_k Q_k B_k^* \tag{5.16}$$

The local estimate $\hat{x}_{(k|k)}^{local}$ is formed by the predicted regional estimate $\hat{x}_{(k|k-1)}^{reg}$ and the local measurement z_k.

$$\hat{x}_{(k|k)}^{local} = \hat{x}_{(k|k-1)}^{reg} + K_k \left(z_k - C_k\, \hat{x}_{(k|k-1)}^{reg} \right) \tag{5.17}$$

The sensor nodes exchange their estimates over the communication channel and combine the estimates in the neighboring nodes N_i.

$$\hat{x}_{(k|k)}^{reg} = \sum_{j \in N_i} W_{ij}\, \hat{x}_{(k|k)}^{Local} \tag{5.18}$$

W is symmetric and its eigenvalues arranges as: $\lambda_1(W) \geq \lambda_2(W)..... \geq \lambda_n(W)$. The second largest eigenvalue $\lambda_2(W)$ is a measure of the speed of consensus algorithm. Thus, the convergence rate of DKF depends on the magnitude of the second largest eigenvalue λ_2. We apply the fast polynomial filter proposed in [16] on the spectrum of W in order to impact the magnitude of $\lambda_2(W)$ that mainly drives the convergence rate. In particular, the convergence is faster when the second largest eigenvalue is small. The polynomial filter of degree m that is applied on the spectrum of W is defined as:

$$p_m(\lambda) = \alpha_0 + \alpha_1\, \lambda + \alpha_2\, \lambda^2 + ... + \alpha_m\, \lambda^m \tag{5.19}$$

The matrix polynomial is given as

$$p_m(W) = \alpha_0\, I + \alpha_1\, W + \alpha_2\, W_2 + ... + \alpha_m\, W_m \tag{5.20}$$

$p_m(W)$ is a periodic update of the current sensor node's value with a linear combination of its previous values. Now we can rewrite Eq. 5.18

$$\hat{x}^{reg}_{(k|k)} = \sum_{i=0}^{m} p_i(W) \left(\hat{x}^{Local}_k \right)_i \tag{5.21}$$

Each sensor node typically applies polynomial filter for distributed consensus. The α_m's are computed off-line assuming that W is known a priori.

The goal is finding the polynomial that leads to the fastest convergence of linear iteration described in Eq. 5.21, for a given weight matrix W and a certain degree m. The optimal polynomial is the one that minimizes the second largest eigenvalue of W. Therefore, we need to solve an optimization problem where the optimization variables are the $m + 1$ polynomial coefficients $\alpha_0, \alpha_1, \ldots \alpha_m$ and the objective function is the spectral radius of $W - \left(\dfrac{1}{n} \right) 11^T$. The following optimization problem needs to be solved:

$$Minimize \; \rho \left(\sum_{i=0}^{m} \alpha_i \; W^i \right), \; where \; \alpha \in \mathbf{R}^{m+1}$$

$$\tag{5.22}$$

$$Subject \; to \; \left(\sum_{i=0}^{m} \alpha_i \; W^i \right) 1 = 1$$

The Linear Matrix Inequality (LMI) of Eq. 5.22 is equivalent to a set of m polynomial inequalities in W, i.e., the leading principal minors of α must be positive. To solve this optimization problem, the auxiliary variable f will be used to bind the objective function, and then the spectral radius constraint is expressed as a linear matrix inequality (LMI). Thus, the following optimization problem needs to be solved.

$$Minimize \; f, \; where \; f \in \mathbf{R}^{m+1}$$

$$Subject \; to \; -fI \le \sum_{i=0}^{m} \alpha_i \; W^i - \frac{11^T}{n} \le fI, \tag{5.23}$$

$$\left(\sum_{i=0}^{m} \alpha_i \; W^i \right) 1 = 1$$

Since W is symmetric, $\sum_{k=0}^{m} \alpha_k \; W^k$ will be symmetric as well. Hence, the constraint $W \, 1 = 1$ is sufficient to ensure that 1 will be also a left eigenvector of W.

Due to the LMI, the above optimization problem becomes equivalent to a semi-definite program (SDP) [16]. SDP is a special case of cone programming and can be efficiently solved by interior point methods. A matrix polynomial p is applied on the weight matrix W to shape its spectrum in order to increase the convergence rate for the DKF. Since the convergence rate is driven by the second eigenvalue $\lambda_2\ (W)$, it is then possible to increase the convergence rate by careful design of the polynomial p. The computation of the coefficients of the optimal polynomial is formulated as a semi-definite program that can be efficiently solved. In addition, the sensors are allowed to use their previous estimates, in order to accelerate the convergence rate in a finite number of steps. Although using the previous estimates will exploit the memory of sensors, the memory requirements can be adjusted to the memory constraints imposed by the sensor.

W is calculated according to the fast polynomial consensus introduced above. Then, each sensor node applies polynomial filter for distributed consensus by implementing the algorithm1. Each sensor uses its previous estimate, in order to accelerate the convergence rate in a finite number of steps.

Algorithm 1 Polynomial Filtered Distributed Consensus

Input: polynomial coefficients α_0, α_1, ..., α_m, tolerance δ

Output: average estimate $\hat{x}^{reg}_{(k|k)}$

While new data exists do

Repeat

  for i = 1,2,3,......, m

$$\hat{x}^{reg}_{(k|k)} = \sum_{i=0}^{m} p_i(W)(\hat{x}^{Local}_k)_i$$

If $\hat{x}^{reg}_{(k|k)} - \hat{x}^{local}_{(k|k)} < \delta$

Exit

else

i = i +1

end if

end

End while

Each node predicts the regional estimate $\hat{x}^{reg}_{(k+1|k)}$ as follow:

$$\hat{x}^{reg}_{(k+1|k)} = A\ \hat{x}^{reg}_{(k|k)} + B_k u_k \tag{5.24}$$

Figure 5.1 presents the proposed architecture of m nodes (in another word, neighbors) running a micro-Kalman filter as a part of the entire n nodes network. It shows also the communication architecture between the nodes. The advantage of the above micro-Kalman filter is that the state estimates produced are identical to the ones obtained via a central Kalman filter, as we will see in the simulation section. Furthermore, a significant advantage for the distributed implementation is the computational

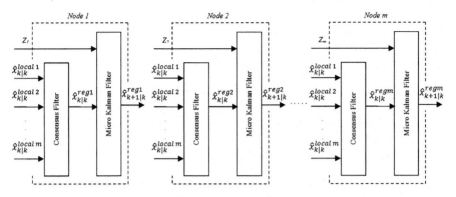

Fig. 5.1 Nodes representation of distributed Kalman filter for m neighbors

costs of the gain matrices. The central Kalman filter gain K has $O\ (m^2n)$ elements while the gain $M\mu$ of the micro-Kalman filter has $O\ (m^2)$ elements. This implies that the implementation of the micro-Kalman filter is more computationally feasible than that of the central Kalman filter, especially for large network. Furthermore, our architecture is scalable in terms of the network size n.

5.5 Simulation Results

In order to compare the performance of the polynomial filter with the standard iterative method. We provide simulation results for different network sizes, where varies from 50 to 150 with step 10. In particular, we are trying to show how fast the polynomial filter is, compared with the standard iteration. We measure the average number of iterations needed by each method to reach the desired level of absolute error across different network sizes with a fixed tolerance $\delta = 10^{-3}$. Figure 5.2 shows that the polynomial filter method is faster than the standard iteration by four times in average. You can notice also that the improvement of polynomial filter methods on the convergence rate is increased in larger networks.

For the sake of completeness, we also provide the output performance of the proposed DKF versus the CKF. Consider a network of $n = 100$ sensor that are distributed randomly with a topology shown in Fig. 5.3. The model of a process is defined as:

$$x_{k+1} = A_k x_k + B_k u_k + w_k \quad k \geq 0 \tag{5.25}$$

$$z_k = C_k x_k + v_k \quad k \geq 0 \tag{5.26}$$

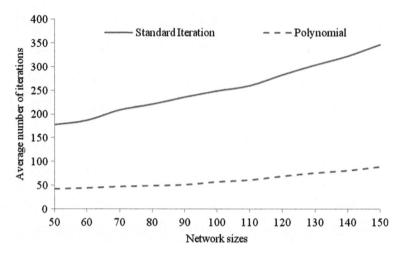

Fig. 5.2 Convergence time for different network sizes

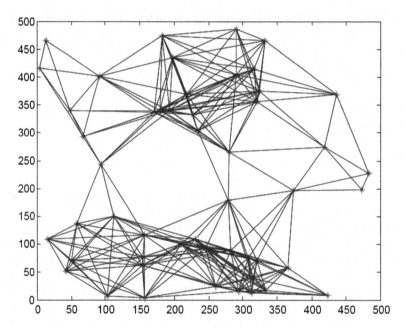

Fig. 5.3 Network topology for n = 100 sensor nodes

The values used for the system defined in Eq. 4.25 are:

$$A = \begin{bmatrix} 0 & -1 \\ 1 & 0 \end{bmatrix}, B = \begin{bmatrix} 1 & 0 \\ 0 & 1 \end{bmatrix}$$

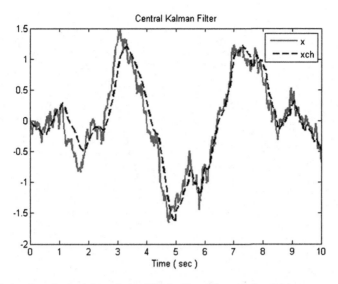

Fig. 5.4 Estimation obtained through the CKF (xch) and the real signal (x)

In the simulations, a heterogeneous network is proposed. Half of the sensors have one kind of sensors and the other half have another kind of sensors (i.e. each half has different C matrix). The two different C matrices used in Eq. 5.26 are:

$$C_1 = \begin{bmatrix} 1 & 0 \\ 0 & 1 \end{bmatrix}$$
$$C_2 = \begin{bmatrix} 1 & 2 \\ 2 & 1 \end{bmatrix}$$

The Simulation time is 10 s with sampling time Ts = 0.01 s and initial value as below:

$$X_0 = \begin{bmatrix} 0 & 0 \end{bmatrix}$$
$$P_0 = \begin{bmatrix} 1 & 0 \\ 0 & 1 \end{bmatrix}$$

The estimation obtained from a Central Kalman Filter CKF, shown in Fig. 5.4, will be our reference to evaluate the proposed DKF performance. Each node in the network has an estimate, Fig. 5.5 shows the squared estimation error for the proposed DKF at node 5 compared with the CKF squared error. Apparently, the proposed distributed and the central Kalman filters provide almost the same estimates, and they can be shown clearly in Fig. 5.6, which shows the average Mean Square Error MSE for all the nodes versus the MSE of the CKF.

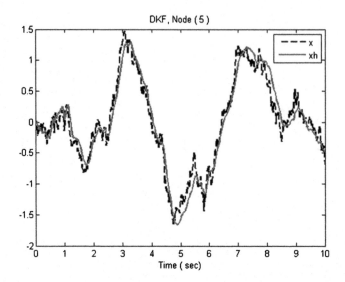

Fig. 5.5 Estimation obtained through DKF (node 5) and the real signal (x)

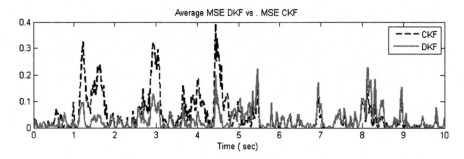

Fig. 5.6 Average MSE for DKF versus MSE for CKF

Olfati in [18] addressed the DKF problem by reducing it into two separate dynamic consensus problems: a low-pass consensus filter for fusion of the measurements and a band-pass consensus filter for fusion of the inverse covariance matrices. He decomposed the central Kalman filter into n micro-Kalman filters with inputs that are provided by two consensus filters. This network of micro-Kalman filters was able to collaboratively provide an estimate of the state of the observed process. Figure 5.7 shows a comparison of the performance of both the proposed distributed Kalman filter and Olfati's algorithm. Simulation results are presented for a wireless sensor network with $n = 200$ nodes and 1,074 links. The result shows that the proposed algorithm improved the average MSE over the olfatis algorithm.

In this work, we use a matrix polynomial p applied on the weight matrix W to shape its spectrum in order to increase the convergence rate. Given the fact that the

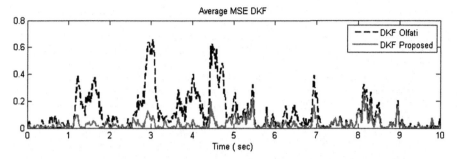

Fig. 5.7 Average MSE for proposed and Olfati's DKF algorithm

convergence rate is driven by the second eigenvalue λ_2 (W), it is then possible to increase the convergence rate by careful design of the polynomial p. We formulate the computation of the coefficients of the optimal polynomial as a semi-definite program that can be efficiently solved.

We also allow the sensors to use their previous estimates, in order to accelerate the convergence rate in a finite number of steps. Although using the previous estimates in our approach will exploit the memory of sensors, the polynomial filter methodology introduced presents three main advantages:

1. It is robust
2. It has explicit control on the convergence rate
3. Its memory requirements can be adjusted to the memory constraints imposed by the sensor.

The proposed polynomial filter increases the convergence rate of the DKF. Fast convergence can contribute to significant energy saving and hence a fast DKF. Multiplication is at the core of DKF operations. Therefore, saving power at the multiplication level will have a significant impact on the energy reserve at each node. Next chapter we propose a light-weight energy-efficient multiplication algorithm for DKF based on Horner's method [19, 20].

Bibliography

1. N.T. John and A. Michael, "Convergence and asymptotic agreement in distributed decision problems," in *Proceeding of the 21st IEEE Conference on Decision and Control*, Orlando, Florida, USA, December 1982, pp. 692–701.
2. R. Wei and R.W. Beard, "Consensus seeking in multiagent systems under dynamically changing interaction topologies," *IEEE Transactions on Automatic Control*, vol. 50, pp. 655–661, November 2005.
3. L. Xiao, S. Boyd, and S. Lall, "A scheme for robust distributed sensor fusion based on average consensus," in *the 4th International Symposium onInformation Processing in Sensor Networks*, pp. 63–70, 2005.

4. C.C. Moallemi and B. Van Roy, "Consensus Propagation," *IEEE Transactions on Information Theory*, vol. 52, pp. 4753–4766, January 2006.
5. D. Spanos, R. Olfati-Saber, and R. Murray, "Distributed sensor fusion using dynamic consensus," in *Proceeding of the 16th IFAC World Congress*, Prague, Czech Republic, July 2005, pp. 199–205.
6. D.S. Scherber and H.C. Papadopoulos, "Locally constructed algorithms for distributed computations in ad-hoc networks," in *Proceeding of the 3rd International Symposium on Information Processing in Sensor Networks*, Berkeley, California, USA, April 2004, pp. 11–19.
7. L. Xiao, S. Boyd, and S.J. Kim, "Distributed average consensus with least-mean-square deviation," in *Proceeding of the 17th International Symposium on Mathematical Theory of Networks and Systems*, Kyoto, Japan, July 2006, pp. 2768–2776.
8. L. Xiao, S. Boyd, and S.-J. Kim, "Distributed average consensus with least–mean–square deviation," *Journal of Parallel and Distributed Computing*, vol. 2, pp. 33–46, May 2007.
9. L. Xiao and S. Boyd, "Fast linear iterations for distributed averaging," *Systems and Control Letters*, pp. 65–78, June 2004.
10. R. Olfati-Saber and R.M. Murray, "Consensus problems in networks of agents with switching topology and time-delays," *IEEE Transactions on Automatic Control*, vol. 49, pp. 1520–1533, January 2004.
11. S. Boyd, A. Ghosh, B. Prabhakar, and D. Shah, "Randomized gossip algorithms," *IEEE Transactions on Information Theory*, vol. 52, pp. 2508–2530, June 2006.
12. S. Sundaram, "Distributed Consensus and Linear Functional Calculation in Networks: An Observability Perspective," in *Proceeding of the 6th International Symposium on Information Processing in Sensor Networks*, Cambridge, Massachusetts, USA, April 2007, pp. 99–108.
13. J. Liu, B.D.O. Anderson, M. Cao, and A.S. Morse, "Analysis of accelerated gossip algorithms," in *Proceeding of the 48th IEEE Conference on Decision and Control*, Chinghai, China, December 2009, pp. 871–876.
14. J. Dong, Q. Chen, and Z. Niu, "Random graph theory based connectivity analysis in wireless sensor networks with Rayleigh fading channels," in *Proceeding of the Asia-Pacific Conference on Communications*, Hong Kong, China, March 2007, pp. 123–126.
15. B. Bollabas, *Random Graphs Second Edition*: Cambride University Press, 2001.
16. E. Kokiopoulou and P. Frossard, "Polynomial Filtering for Fast Convergence in Distributed Consensus," *IEEE Transactions on Signal Processing*, vol. 57, pp. 342–354, October 2009.
17. S. Kar and J.M.F. Moura, "Sensor Networks With Random Links: Topology Design for Distributed Consensus," *IEEE Transactions on Signal Processing*, vol. 56, pp. 3315–3326, March 2008.
18. R. Olfati-Saber, "Distributed Kalman Filter with Embedded Consensus Filters," in *the 44th IEEE Conference on Decision and Control*, pp. 8179–8184, 2005.
19. A. Abdelgawad and M. Bayoumi, "Low Power Distributed Kalman Filter for Wireless Sensor Networks," *EURASIP Journal on Embedded Systems*, vol. 2011, Article ID 693150, 11 pages, doi:10.1155/2011/693150, 2011.
20. A. Abdelgawad and M. Bayoumi," Distributed Kalman Filter Using Fast Polynomial Filter," *IEEE International Symposium on Circuits and Systems, ISCAS 2011*, 15–18 May 2011

Chapter 6
Proposed Multiplication Algorithm for DKF

Abstract An efficient and low-power multiplication algorithm has been proposed in this chapter. It reduces the number of add operations during multiplication by rounding any sequence of 1s in the fractional part. The impact of using the proposed multiplication method on FIR and IIR filters response has been studied. Experimental results show that the proposed algorithm achieves up to 17% power saving and 16% increasing in speed, with only 1% accuracy loss compared to Horner's algorithm. The new multiplication method has been validated experimentally using the eZ430-RF2500 wireless sensor board. In the next chapter, we will study the impact of using the proposed multiplication method on the power consumption of the proposed DKF.

6.1 Introduction

Signal processing in wireless sensor network has a vast range of applications. Finite Impulse Response filter (FIR), Infinite Impulse Response filter (IIR), and Kalman Filter find applications in object tracking, environmental monitoring and many other applications. These tasks are computationally intensive and strain the energy resources of any single computational node in a wireless sensor network. In other words, most sensor nodes do not have the computational resources to complete many of these signal processing tasks repeatedly. Multiplication is at the core of many data and signal processing operations. Therefore, saving power at the multiplication level has a significant impact on the energy reserve of each node. Consequently, energy-efficient multiplication can extend the WSN's lifetime and increase its computational capabilities per Watt of power. The sensor nodes available in the market, such as the Crossbow's Micaz [1] and Telosb [2] motes, depend on an 8-bit (in Micaz) or 16-bit (in Telosb) microcontroller. These microcontrollers do not have a floating-point multiplier. Moreover, their hardware multiplier, if enabled, tends to deplete the nodes' energy quickly. To deal with such systems,

A. Abdelgawad and M. Bayoumi, *Resource-Aware Data Fusion Algorithms*
for Wireless Sensor Networks, Lecture Notes in Electrical Engineering 118,
DOI 10.1007/978-1-4614-1350-9_6, © Springer Science+Business Media, LLC 2012

several multiplication algorithms have been proposed which rely on repeated additions and consume lots of instruction cycles and exhibits limited precision.

In this work we propose a light-weight energy-efficient multiplication algorithm based on Horner's method. Our method aims to reduce the number of add operations during multiplication by rounding any sequence of 1s in the fractional part. The applied rounding reduces the number of instruction cycles, and reduces the memory storage without increasing the code complexity or sacrificing accuracy.

6.2   Overview of Multiplication Algorithms

Multiplications are often implemented with shift- and-add operations for hardware efficiency [3]. In this method, a set of partial products is formed by multiplying the multiplicand by each digit of the multiplier. Each partial product is shifted one digit position from the previous partial product, and the partial products are then added to produce the final product. Binary multiplication is done the same way; however, because binary numbers consist only of 1s and 0s, each partial product will be either an exact copy of the multiplicand, or it will be zero. Those bit positions of the multiplier which contain 1's produce partial products equal to the multiplicand; those bit positions of the multiplier which contain 0s produce partial products which are equal to zero. As an example, consider the multiplication of the two numbers A and B below, represented in 12-bits.

A = 0.14325 = 0.001001001010b
B = 0.12345 = 0.000111111001b

The traditional method to perform this multiplication is: $0.14325 * 0.12345 +$ $0.001001001010b * (2^{-4}+2^{-5}+2^{-6}+ 2^{-7}+ 2^{-8}+2^{-9}+2^{-12})$

$\quad$ = 0.000000100100b +
$\quad$ 0.000000010010b +
$\quad$ 0.000000001001b +
$\quad$ 0.000000000100b +
$\quad$ 0.000000000010b +
$\quad$ 0.000000000001b +
$\quad$ 0.000000000000b +
$\quad$ 0.000001000110b $\quad$ = 0.01708984375

The exact result of this multiplication is 0.0176842125. The traditional method results in an absolute error of 0.00059436875, which is approximately 2.5 LSB. This error can be attributed to finite word length effects due to register width limitations. As the number of bits allocated for the fractions increases, this error is reduced. The Horner's method aims to reduce this error while maintaining the same register widths.

Horner's method is primarily designed to perform multiplication on devices that do not have a dedicated hardware multiplier [4]. It dictates a set of design equations, which are unique for any multiplier. These design equations directly relate to a sequence of shift and add operations on the multiplicand. The Horner's algorithm is based on the positions of the 1s in the multiplier and their distance to the immediate 1 to their left. This is done starting from the rightmost bit position and moving left until the last 1 before the binary point. In the binary equivalent of the multiplier $0.14325 = 0.001001001010_b$, starting from the right, the first 1 occurs at bit position 2^{-11}. The difference in position of this 1 to its immediate 1 to the left is two. Similarly, the difference for the 1 in bit position 2^{-9} is three and so on. If the number to be multiplied is denoted as A, the design equations can be written as:

1. $A1 = A * 2^{-3} + A$: Set the intermediate result equal to the operand B and start with the rightmost 1. For the first iteration, the weight 2^{-3} is applied to the intermediate result as the distance of the rightmost 1 (bit position 2^{-12}) in the multiplier to its next 1 (bit position 2^{-9}) is three.

$$0.000001001001_b +$$
$$0.001001001010_b$$
$A1 = 0.001010010011_b$

2. $A2 = A1 * 2^{-1} + A$: Continue to the next 1 in bit position 2^{-9}.The weight 2^{-1} is now applied to the intermediate result since the distance of the 1 in bit position 2^{-9} to its next 1 (bit position 2^{-8}) is one. The operand is again added.

$$0.000101001001_b +$$
$$0.001001001010_b$$
$A2 = 0.001110010011_b$

3. $A3 = A2 * 2^{-1} + A$: Keep on to the next 1 in bit position 2^{-7} The weight 2^{-1} is applied to the intermediate result and the operand added.

$$0.000111001001_b +$$
$$0.001001001010_b$$
$A3 = 0.010000010011_b$

4. $A4 = A3 * 2^{-1} + A$: Go on to the next 1 in bit position 2^{-6} The weight 2^{-1} is applied to the intermediate result and the operand added.

$$0.001000001001_b +$$
$$0.001001001010_b$$
$A4 = 0.010001010011_b$

5. $A5 = A4 * 2^{-1} + A$: Continue to the next 1 in bit position 2^{-5} The weight 2^{-1} is applied to the intermediate result and the operand added.

$$0.001000101001_b +$$
$$0.001001001010_b$$
$A5 = 0.010001110011_b$

6. $A6 = A5 * 2^{-1} + A$: Keep on to the next 1 in bit position 2^{-4} The weight 2^{-1} is applied to the intermediate result and the operand added.

$$0.001000111001_b +$$
$$0.001001001010_b$$
$A6 = 0.010010000011_b$

The result $= A6 * 2^{-4}$ continues to the last 1 in bit position 2^{-4}. The factor 2^{-4} is applied to the intermediate result, as it is the weight at the position of the leftmost 1. The operand is not added this time, since all the 1s have been taken into account. The result $= A6 * 2^{-4} = 0.000001001000_b = 0.017578125$. This has an absolute error of 0.0001060875 which is just 0.434534 LSB, which is 0.60% error from the actual result.

6.3 Proposed Method

The proposed method is a method targeted for fixed-point multiplication by utilizing the redundancy of signed digit code. The feature of redundancy in this representation allows a coefficient implementation to be selected which in general requires fewer additions and thus yields a faster compact multiplication. The proposed method aims to reduce the number of add operations during multiplication by rounding any sequence of 1s in the fractional part. For example the number 1010.010111101 becomes 1010.01100001.

To better illustrate the algorithm, consider the following example where the number 0.12345 is multiplied by the constant 0.14325.

$A = 0.14325 = 0.001001001010b$
$B = 0.12345 = 0.000111111001b$

The multiplicand B is rounded according to the proposed method to yield B_{new}.
$B_{new} = 0.12345 = 0.001000000001_b$
The algorithm, then, follows Horner's method but with only two steps compared to Horner's seven steps presented in the previous section.

1. $A1 = A * 2^{-9} + A$: Set the intermediate result equal to the operand B_{new} and start with the rightmost 1. For the first iteration, the weight 2^{-9} is applied to the intermediate result as the distance of the rightmost 1 (bit position 2^{-12}) in the multiplier to its next 1 (bit position 2^{-3}) is 3.

$$0.000000000001_b +$$
$$0.001001001010_b$$
$$\overline{A1 = 0.001001001011_b}$$

The result $= A1 * 2^{-3}$: Proceed to the last 1 in bit position 2^{-3}. The factor 2^{-3} is applied to the intermediate result, as it is the weight at the position of the leftmost 1. The operand is not added this time, since all the 1s have been taken into account. The result $= A1 * 2^{-3} = 0.000001001001_b = 0.017822265625$. This has an absolute error of 0.000138053125 which is just 0.5654656 LSB, which is 0.78% error from the actual result. The procedure remains the same if the operand is a negative fraction.

6.4 Simulation Result

The error of the multiplication comes from the fraction part and depends on the total number of bits in the fractional part. Matlab was used to compare the accuracy for the proposed method with the existing methods; we calculated the average absolute error of multiplying two fractions for different fraction width (starting from 2 bits to 12 bits). We multiplied all the possible combinations of the two fractions and got the absolute average error. Figure 6.1 shows the average absolute error for the shift-and-add, Horner, and proposed methods. The absolute average error of the proposed method is very close to Horner's method. The simulation results show that

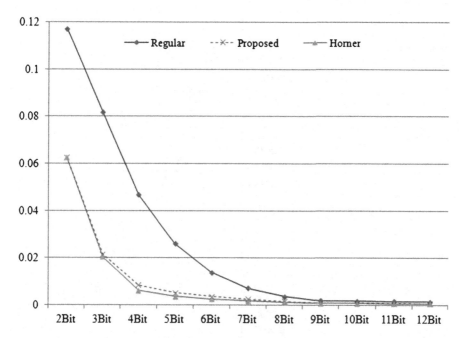

Fig. 6.1 Absolute average multiplication error for both methods

the proposed method reduces the accuracy by a maximum of 1% compared to Horner's method. Figure 6.2 draws a box and whisker diagram to show the spread of the absolute error of the proposed multiplication method.

Table 6.1 shows the comparison of speed, accuracy and memory requirements for both methods. The proposed method reduces the number of instruction cycles and the code size without scarifying the accuracy.

6.5   Case Study

In this section, we are studying the impact of using the proposed multiplication method on FIR and IIR filters response. The basic operation needed to implement a FIR [5] filter is the multiply-and-accumulate (MAC). The mathematical expression for the FIR filter is:

$$Y(k) = \sum_{i=0}^{n} c_i * X(n - i) \tag{6.1}$$

where k is the time step, $Y(k)$ is the filter output at time k, $X(n-i)$ is the sampled input at time $n-i$, c_i is the filter coefficient i, and N is the order of the filter.

Consider a low pass FIR filter of order 12 with the following coefficients [0.0002, −0.0024, −0.0158, −0.0190, 0.0723, 0.2714, 0.3867, 0.2714, 0.0723,

Fig. 6.2 Box-and-whisker diagram of the proposed multiplication error

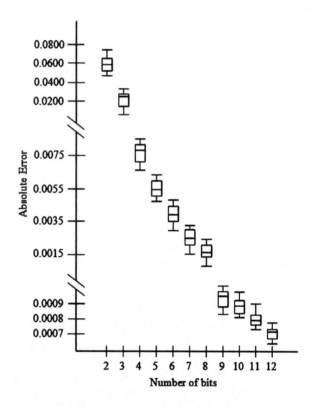

Table 6.1 Comparison of speed, accuracy and memory requirements for both methods

Type	Method	Instruction cycle	Code size	Result	Absolute error
Integer–float multiplication	Horner	32	60	18,115	0.3375
41 * 441.8375	Proposed	29	60	18,115	0.3375
Float–float multiplication	Horner	18	37	0.0175781	0.000106
0.14325 * 0.12345	Proposed	6	26	0.0178222	0.000138

−0.0190, −0.0158, −0.0024, and 0.0002]. Figure 6.3 shows the magnitude and the phase response of the FIR filter using Horner's method and the proposed multiplication method.

The IIR filter normally includes adders and multipliers working at a very high speed; it is important to design fast Multipliers [6]. The IIR filter is represented by a difference equation where the output signal at a given instant is obtained as a linear combination of samples of the input and output signals at previous time instants. Moreover, an instantaneous dependency of the output on the input is also usually included in the IIR filter. The difference equation that represents an IIR filter is:

$$Y(n) = \sum_{i=0}^{n} b_i * X(n-i) - \sum_{i=0}^{m} a_i * X(n-i) \qquad (6.2)$$

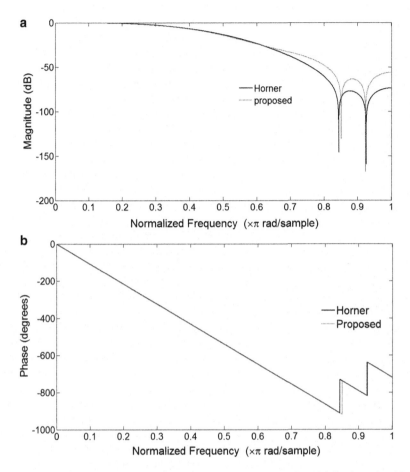

Fig. 6.3 FIR filter response using the Horner and the proposed multiplication algorithms. (a) Magnitude response, (b) phase response

For an IIR filter, coefficients refer to the (n * 1) vector a and (m * 1) vector b. Consider a high pass IIR filter of order 12 with the following coefficients {b = [1 −1.9082 1] and a = [−1.0644 0.8125]}. Figure 6.4 shows the magnitude and the phase response of the IIR filter using Horner and the proposed multiplication algorithms. Figures 6.3 and 6.4 Show that the proposed method does not affect the response of both IIR and FIR filters.

6.6   Counter Example Power Measurement

For demonstration purposes, we considered the MSP430 microcontroller. The MSP430 is a family of ultra-low power microcontrollers by Texas Instruments [7]. Low-cost, low power and a powerful instruction set make MSP430 an ideal

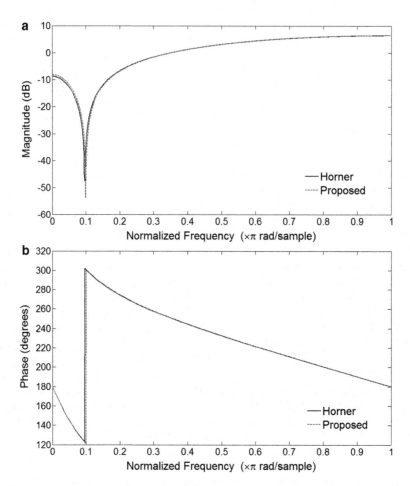

Fig. 6.4 IIR filter response using the Horner and the proposed multiplication algorithms. (**a**) Magnitude response, (**b**) phase response

choice microcontroller for WSNs. The MSP430 microcontroller CPU can perform a register shift or add in one instruction cycle. This allows fast execution of multiplications using the proposed method.

In order to compare between the power consumption of the proposed method and the existing methods, a case of multiplying two fractions (0.14325 * 0.12345) is implemented on MSP430F2274 microcontroller (eZ430-RF2500 kit) using IAR Embedded Workbench Ver. 3.41A. Experimental results show that the proposed method reduces the current by 15 nA and increases the speed by 16 ns with only 0.18% accuracy loss as shown in Fig. 6.5. The proposed method, in the best case, achieves up to 17% power saving and 16% increase in speed, with only 1% accuracy loss compared to Horner's algorithm [8].

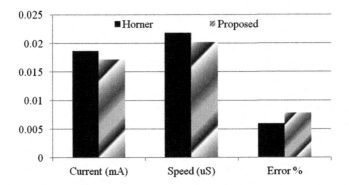

Fig. 6.5 Current, speed, and error both methods

In order to implement this DKF in wireless sensor network the computation issue will come up. Moreover, Kalman filter is a power hungry algorithm in term of computational complexity.

In this chapter we offered a solutions for this problem; we proposed a novel light-weight and low-power multiplication algorithm. The proposed algorithm aims to decrease the number of instruction cycles, save power and reduce the memory storage without increasing the code complexity or sacrificing accuracy. More experimental results for the proposed DKF using the proposed multiplication algorithm will be presented chapter.

Bibliography

1. Crossbow Technology, "Micaz datasheet," http://www.xbow.com/Products/Product_pdf_files/Wireless_pdf/MICAZ_Datasheet.pdf.
2. J. Polastre, R. Szewczyk, and D. Culler, "Telos: enabling ultra-low power wireless research," in *Proceeding of the Information Processing in Sensor Networks*, pp. 364–369, November 2005.
3. H.T. Nguyen and A. Chattejee, "Number-splitting with shift-and-add decomposition for power and hardware optimization in linear DSP synthesis," *IEEE Transactions on Very Large Scale Integration Systems,* vol. 8, pp. 419–424, May 2000.
4. K. Venkat and M. Raju, "Efficient Signal Conditioning for Microcontroller Based Medical Solutions," in *Proceeding of the IEEE International Symposium on Consumer Electronics*, Dallas, Texas, USA, June 2007, pp. 1–5.
5. R. Tamura, M. Honma, N. Togawa, M. Yanagisawa, T. Ohtsuki, and M. Satoh, "FIR filter design on Flexible Engine/Generic ALU array and its dedicated synthesis algorithm," in *Proceeding of the IEEE Asia Pacific Conference on Circuits and Systems*, Macao, China, December 2008, pp. 701–704.
6. R. Landry, Jr., V. Calmettes, and E. Robin, "High speed IIR filter for XILINX FPGA," in *Proceeding of the Midwest Symposium on Circuits and Systems,* Notre Dame, Indian, USA, August 1998, pp. 46–49.
7. Texas Instruments Inc., "MSP430 family of microcontrollers," http://www.ti.com/msp430.
8. A. Abdelgawad, S. Abdelhak, S. Ghosh, and M. Bayoumi, "A low-power multiplication algorithm for signal processing in wireless sensor networks," in *Proceeding of the 52nd IEEE International Midwest Symposium on Circuits and Systems*, Cancun, Mexico, August 2009, pp. 695–698.

Chapter 7
Experimental Results for the Proposed DKF

Abstract Experimentally, the proposed DKF using the proposed multiplication method and the proposed fast polynomial filter was evaluated. The DKF introduced by Olfati was experimentally tested as well. The results show that the proposed DKF achieves up to 33% energy saving. The results show also that one node can run the Olfati's DKF for up to five neighbors only, but the proposed DKF can run for up to seven neighbors. This different in the nodes numbers is because of the memory limitation, as Olfati's DKF exchange the measurements and the covariance, but the proposed DKF exchange the estimation only. Moreover the proposed multiplication method saves memory as well.

7.1 Test Bed

A test bed composed of 20 wireless sensor motes – TelosB – was used to test the proposed DKF and measure its power consumption. TelosB is designed for low-power operation. The low power operation of the TelosB module is due to the ultra low power Texas Instruments MSP430 F1611 microcontroller featuring 10 kB of RAM, 48 kB of flash, and 128 B of information storage. The MSP430 microcontroller is based on a 16-bit RISC core integrated with RAM and flash memories, analog and digital peripherals and a flexible clock subsystem. It supports several low-power operating modes and consumes as low as 1 µA in a standby mode; it also has very fast wake up time of no more than 6 µs. TelosB features a Chipcon 2420 radio in the 2.4 GHz band. The CC2240 is controlled by the MSP430 microcontroller through the SPI port and a series of digital I/O lines with interrupt capabilities. The MAC protocol used is X-MAC. X-MAC is an asynchronous MAC protocol in which the sender uses short preambles to awaken the receiver. Before any transmission, the sender senses the channel; if it is busy the sender retries after a random backoff, otherwise it sends short preambles embedding the address of the receiver. Once the receiver detects its address, it sends an acknowledgment, and the sender can start transmitting the data [1].

A. Abdelgawad and M. Bayoumi, *Resource-Aware Data Fusion Algorithms*   101
for Wireless Sensor Networks, Lecture Notes in Electrical Engineering 118,
DOI 10.1007/978-1-4614-1350-9_7, © Springer Science+Business Media, LLC 2012

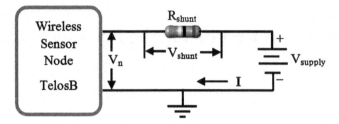

Fig. 7.1 Power measurement with shunt resistor

Energy consumption in each TelosB can be attributed to the current draw of each node. Therefore, we can use accurate measurements of the amount of current that the node sinks to determine the power consumption. Current measurement is typically done with a shunt resistor placed in series with the current flow in a circuit as shown in Fig. 7.1. This resistor is specifically chosen to be high-precision and low-impedance so as not to interfere greatly with the circuit being monitored. Because the value of the resistor is known, by measuring the voltage drop across the shunt resistor, we can accurately calculate the current using Ohm's law as in Eq. 5.3.

$$P = V_n * I = \left(V_{supply} - V_{shunt}\right) * \frac{V_{shunt}}{R_{shunt}} \tag{7.1}$$

The code for all the nodes was written in NesC. A java GUI was developed to provide a friendly user interface with the nodes. The java interface is a multi-threaded socket-based program that communicates with a serial forwarding program. A set of 20 nodes is distributed around the laboratory, and a laptop gateway is configured to be able to send/receive control signals and data packets to/from the nodes [2].

7.2 Experimental Results

To illustrate the effects of energy saving for the proposed multiplication method on the DKF, Fig. 7.2 shows the comparison between the power consumption of one node has five neighbors and run DKF using the proposed multiplication method and Horner's method. Figure 7.2 shows the power trace for only one iteration. The proposed method takes 140 ms while the Horner's method takes 153 ms. Thus, using the proposed multiplication method in DKF saves 8% of energy.

The proposed polynomial filter increases the convergence rate of the DKF. Fast convergence can contribute to significant energy saving and hence a fast DKF. Table 7.1 shows the time and energy consumption for the DKF using the standard polynomial and the proposed polynomial. The measurements for one node have five neighbors and runs for ten iterations.

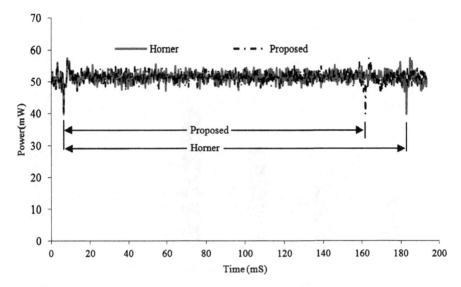

Fig. 7.2 Power traces for DKF using proposed and Horner multiplication methods

Table 7.1 Energy and time for the proposed polynomial filter		Proposed polynomial	Standard polynomial
	Energy (mJ)	62.01644	71.05673
	Time (S)	14.6193	16.5402

Experimentally, the proposed DKF using the proposed multiplication method and the proposed fast polynomial filter was evaluated. The DKF introduced by Olfati was experimentally tested as well. Figure 7.3 shows the comparison for both methods for different numbers of neighbors. The results show that the proposed DKF achieves up to 33% energy saving. The results show also that one node can run the Olfati's DKF for up to five neighbors only, but the proposed DKF can run for up to seven neighbors. This difference in the nodes numbers is because of the memory limitation, as Olfati's DKF exchange the measurements and the covariance, but the proposed DKS exchange the estimation only. Moreover, the proposed multiplication method saves memory as well [2].

We have presented a low power distributed Kalman filter based on a fast polynomial filter [3]. Fast convergence led to significant energy saving. In addition, we proposed a light-weight energy-efficient multiplication algorithm. The proposed multiplication method reduced the number of add operations during multiplication by rounding any sequence of 1s in the fractional part. The applied rounding reduced the number of instruction cycles, and reduced the memory storage without increasing the code complexity. The experimental results show that the proposed DKF achieved up to 33% energy consumption save compared to Olfsti's DKF. Moreover, the proposed DKF efficiently uses the node's memory, so each node can run DKF with up to seven neighbors.

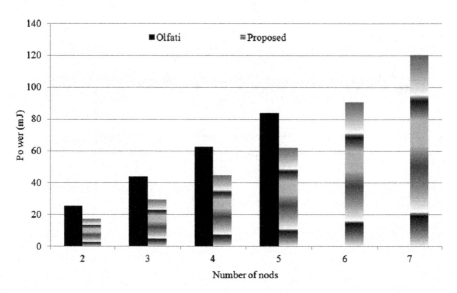

Fig. 7.3 Energy consumption of the proposed DKF and Olfatis' DKF

Bibliography

1. E.A.M. Buettner, G. Yee, and R. Han, "X-mac: A short preamble mac protocol for duty-cycled wireless sensor networks," in *Proceeding of the 4th ACM Conference on Embedded Sensor Systems*, New York, NY, USA, April 2006, pp. 307–320.
2. A. Abdelgawad, S. Abdelhak, S. Ghosh, and M. Bayoumi, "A low-power multiplication algorithm for signal processing in wireless sensor networks," in *Proceeding of the 52nd IEEE International Midwest Symposium on Circuits and Systems*, Cancun, Mexico, August 2009, pp. 695–698.
3. A. Abdelgawad and M. Bayoumi, "Low Power Distributed Kalman Filter for Wireless Sensor Networks," *EURASIP Journal on Embedded Systems*, vol. 2011, Article ID 693150, 11 pages, doi:10.1155/2011/693150, 2011.

Index

A. Abdelgawad and M. Bayoumi, *Resource-Aware Data Fusion Algorithms*
for Wireless Sensor Networks, Lecture Notes in Electrical Engineering 118,
DOI 10.1007/978-1-4614-1350-9, © Springer Science+Business Media, LLC 2012